William Lintern

The mineral surveyor & valuer's complete guide

Comprising a treatise on improved mining surveying

William Lintern

The mineral surveyor & valuer's complete guide
Comprising a treatise on improved mining surveying

ISBN/EAN: 9783337203948

Printed in Europe, USA, Canada, Australia, Japan

Cover: Foto ©berggeist007 / pixelio.de

More available books at **www.hansebooks.com**

CROSBY LOCKWOOD & SON, 7, Stationers' Hall Court, E.C.

MECHANICAL ENGINEERING, &c.—contd.

Sewing Machinery. J. W. Urquhart	2/–
Power of Water. J. Glynn	2/–
Power in Motion. J. Armour	2/–
Iron and Heat. J. Armour	2/6
Mechanism and Machines. T. Baker & J. Nasmyth	2/6
Mechanics. C. Tomlinson	1/6
Cranes and Machinery. J. Glynn	1/6
Smithy and Forge. W. J. E. Crane . .	2/6
Sheet-Metal Worker's Guide. W. J. E. Crane . .	1/6

MINING & METALLURGY.

Mineralogy. A. Ramsay	3/6
Coal and Coal Mining. Sir W. W. Smyth . . .	3/6
Metallurgy of Iron. H. Bauerman	5/–
Mineral Surveyor's Guide. W. Lintern . . .	3/6
Slate and Slate Quarrying. D. C. Davies . .	3/–
Mining and Quarrying. J. H. Collins . . .	1/6
Subterraneous Surveying. T. Fenwick & T. Baker	2/6
Mining Tools. W. Morgans	2/6
Plates to ditto. 4to.	4/6
Physical Geology. Portlock & Tate . . .	2/–
Historical Geology. R. Tate	2/6
The above 2 vols., bound together	4/6
Electro-Metallurgy. A. Watt	3/6

NAVIGATION, SHIPBUILDING, &c.

Navigation. J. Greenwood & W. H. Rosser . .	2/6
Practical Navigation. Greenwood, Rosser & Law .	7/–
Navigation and Nautical Astronomy. J. R. Young	2/6
Mathematical & Nautical Tables. Law & Young .	4/-
Masting and Rigging. R. Kipping	2/–
Sails and Sailmaking. R. Kipping	2/6
Marine Engines. R. Murray & G. Carlisle . .	4/6
Iron Ship-Building. J. Grantham	4/-
Naval Architecture. J. Peake	3/6
Ships, Construction of. H. A. Sommerfeldt . .	1/6
Plates to ditto, 4to	7/6
Ships and Boats. W. Bland	1/6

CROSBY LOCKWOOD & SON, 7, Stationers' Hall Court, E.C.

WEALE'S SCIENTIFIC & TECHNICAL SERIES.

AGRICULTURE & GARDENING.

CROSBY LOCKWOOD & SON, 7, Stationers' Hall Court, E.C.

THE

MINERAL SURVEYOR & VALUER'S
COMPLETE GUIDE

COMPRISING A TREATISE ON IMPROVED MINING
SURVEYING AND THE VALUATION OF
MINING PROPERTIES

WITH

NEW TRAVERSE TABLES

By WILLIAM LINTERN
MINING AND CIVIL ENGINEER

FOURTH EDITION, WITH AN APPENDIX ON
MAGNETIC AND ANGULAR SURVEYING

WITH RECORDS OF THE PECULIARITIES
OF NEEDLE DISTURBANCES

With Four Plates of Diagrams, Plans, &c.

LONDON
CROSBY LOCKWOOD AND SON
7, STATIONERS' HALL COURT, LUDGATE HILL
1898

LONDON :
PRINTED BY J. S. VIRTUE AND CO , LIMITED,
CITY ROAD.

PREFACE.

THE TRAVERSE TABLES herewith submitted to the public were computed some years ago by the author, to supply a need he had often felt in his professional duties, when engaged in laying out works for the connection of different and distant operations in collieries and mines: viz. a Set of Tables by the use of which the results of some extensive surveys may be computed, and acted upon, with greater accuracy and confidence than can be sometimes attained by the ordinary methods of plotting.

It occurred to the author that if a survey were made with great care and accuracy, the results should be capable of computation with absolute certainty; and that thus much cost in mining operations would occasionally be saved, by the drivages meeting properly, as intended: for where the cost of driving is large, in hard ground for instance, frequently costing £3, or £4, or more per yard, an error even of 5 yards may thus become an expensive matter; and frequently also there will be no inconsiderable expense in endeavouring to "straighten" a crooked driving, which has resulted from incorrect "points" having been given.

. It is not claimed for these Tables that they will supersede good surveying; on the contrary, if the surveying has been incorrectly done the Tables will not, and cannot, correct the error; but where the work of surveying has been done correctly, the employment of these Tables will eliminate the result much more unerringly than plotting; and thus they have often been proved by the author to be very useful and valuable.

Their application to surveys of mineral workings or openings which "pitch" highly must also be a matter of great utility, and they will supply a very ready medium, by which to ascertain the horizontal measure of the different pitches of the lines taken, which often vary very much in the same mine or colliery.

The use of the Magnetic needle is herein strongly defended, as being more generally applicable, as a rule, to mining surveying than any other system that can be suggested. But the author takes the liberty of submitting that the *magnetic meridian* is not a satisfactory "base" of working: and a description will be found herein of an improved circumferentor, by the use of which the True Meridian may be made the regular working "base," instead of the Magnetic Meridian; and by which the error occasioned by the constantly varying declination of the latter may be effectually guarded against.

A description is also given herewith of an improved protractor, by which the angles may be taken off directly from the instrument itself, instead of from points made in the paper either by the "folding arms" or pencil employed with the plainer kinds: this the author thinks must facilitate the making of a good and correct plan, and save the paper from those numerous punctures which are made in using the existing kinds of protractors. .

A general description of the operations of careful mining surveying will also be found in the chapters preceding the Tables; and some of the peculiarities, both of home and foreign mining, are there pointed out for the benefit of those who may have to engage therein, and prepare plans, etc. of the same.

A chapter is also given on the Valuation of mining properties, and some examples of the varying class of property connected with mining, which may be also occasionally useful.

The work of surveying etc. enters largely into mining, and the cost of mining is sometimes affected in no slight degree by the manner in which the work of the surveyor is accomplished: and he should therefore omit no opportunity of performing his task in a manner to yield the best possible results. If the accompanying matter and Tables will in any manner facilitate the attainment of these desiderata the author will be pleased, and his anticipations realized.

Possibly some errors may hereafter be discovered in the work, but as every care has been taken to ensure accuracy, it is hoped they will not be found to be numerous.

If any justification is considered necessary for having written this treatise with the Tables for publication, it is submitted in these few words:—the author does not doubt but that others have felt the want that he himself felt, before he set about the computation of these Tables; and that there are probably many who would like to avail themselves of such assistance as they may afford, at a less cost than that of having to construct them for themselves. Of the text it may be said that it seems to be but a natural accompaniment of the Tables.

PREFACE TO THE SECOND EDITION.

THE subject of Magnetic Surveying having been a matter of inquiry with some persons who had purchased the first edition of this work; and there being a disposition with some people to depreciate the value of the magnetic needle as an instrument for surveying, accounting it of inferior practical value for the purposes indicated; I have been induced to add to this edition the leading results of a not inconsiderable experience in this field of labour, and the conclusions arrived at from a (perhaps) more than general close observation of the behaviour of the magnetic needle, under considerably varying circumstances.

That the magnetic needle is a very valuable instrument for certain operations in mining surveying there can, I think, be no doubt; and my object has been to show the class of operations in which it may be employed with advantage, as well as to indicate those which are more suited to a different method of surveying. It is hoped that the description and examples given will assist the thoughtful student to form an intelligent judgment in the matter treated of, and enable him to come to a right conclusion as to which system of working will be the most suitable to any particular class of work he may have in hand.

A few notes on colliery operations, &c., are added at the end, which it is hoped may be read with interest.

July, 1887.

CONTENTS.

APPENDIX.

CHAPTER I.

PAPER FOR MAPS.

WHEN a Survey of a property is about to be made for the purpose of obtaining a correct plan, it is important to secure a sheet of good paper mounted on cloth, of about the necessary size, and to keep it in a dry and moderately cool place for some few days before anything is done towards laying down any part of the survey; if this were not attended to, but the paper ordered from the maker and the plotting proceeded with immediately on its arrival, it would soon be discovered that lines which were laid down with care at the first according to scale, shew a different length on the scale being applied to them some little time after, supposing the paper to have remained in a dry room ; and it would be a hopeless task to endeavour to get lines laid down subsequently, to "come in" correctly with those laid down at the first. Paper arriving fresh from the mill is sure to be more or less damp, and after it has been exposed a little time to a dry atmosphere the damp evaporates, and, as a natural consequence, the paper shrinks ; and until ample time has been allowed for this evaporation, in a medium dry and warm room, the paper should not be used, but should be occasionally unrolled and turned. It will also sometimes occur that paper which has been at hand a considerable time, has been kept in some very dry cupboard, or place in which it may have shrunk below its ordinary dimensions : in this case, if lines were to be laid down to scale immediately on bringing it out, and the paper then left exposed to ordinary atmospheric influences, it would be found to have expanded, and on applying the scale to the lines previously laid down, it would be found that they now represented a greater length than was originally intended.

Some persons may think that these are unnecessary precautions, and that no appreciable error or difference arises from such causes, but such is far from being the case; when a plan is being made to, say, 2 chains to an inch, and the lines laid down are some of them from 20 to 30 chains each, it would then be no unusual thing to find a difference of 20 links and upwards in the longest lines, and this represents an error, multiplied over two or three such lines, sufficient in some cases to give rise to disagreeable consequences, and to law suits occasionally for infringements of boundaries, the costs of which are sometimes heavy.

This is only one of the many little points of detail which go to make up the difference between a good plan and a bad one,—between an accurate plan and an inaccurate; for with a material so absorbent of moisture as paper is, it is most natural to expect that in laying down long lines to scale, without giving any attention to the normal, or abnormal state of the paper, differences and errors are sure to arise and to accumulate.

Where the plan to be made is a permanent working plan the best paper should be procured, for inferior papers wear badly, and are easily chafed off with the action of indiarubber in the operation of cleansing them from the dust and pencil marks, which are the inevitable accompaniments of a mining plan; but an inordinate use of indiarubber only goes to destroy the face of the paper, and causes it to become the more easily soiled.

The best hot-pressed paper is considered the most durable, and best to work upon; and it should be well mounted either on white or brown cloth, and bound at the edges.

CHAPTER II.

DIVIDING AND LINING THE PAPER.

HAVING given the paper sufficient time to season properly, the next thing that should be done with it is to carefully lay down two series of lines, dividing the paper into squares of, say, 10 chains each way, or 10 acres area each; one set of these lines should represent a meridian (which of the meridians it should be we will hereafter discuss), and the other set should traverse the first at right angles. This operation should be performed with scrupulous care, and to do it properly and accurately a large beam compass is necessary, and a long thin true straight-edge; if one of steel is at hand so much the better,—there are few things more useful and valuable to the surveyor.

If a hand compass (or ordinary dividers) is used to mark off on the two base lines (which a person of experience would at once recognise as the first step) traversing each other, it should be so set that it will accurately divide such a sufficient number of tens, taken together, as shall ensure the accuracy of each division; since it is, in a matter of this kind, unwise to trust to the setting of the compass to one single measure on the scale:—for instance, if a divided beam compass is used the hand compass should be passed over such a number of the multiples of the division determined on for the side of the squares on the paper, as shall ensure its dividing accurately over the length and breadth of the paper.

The lines should then be finely and truly drawn through the points so marked with faint red ink, or some other colour that will be sufficiently distinguishable from the other lines to be drawn upon the plan.

It will be necessary to determine, before the lines are laid down, what direction on the paper the meridional lines must take so as to ensure the plan of the property being got fairly

into the body of the paper, or else the conformation of the
outline of the boundary to the edges of the paper will never
have a pleasing effect on the eye; this may usually be
ascertained from some old plan, or by taking the bearing of
some distant object with the theodolite from one of the
angles of the boundary, or a combination of two or three such
lines across the middle of the property. The correct laying
down of these sets of lines is a matter of great importance, as
instances are frequently met with where plans of properties of
300 or 400 acres area, or more are made, and the magnetic
meridian is represented by a short line in one corner of the
paper of about 14 or 16 inches in length; and to plot a series
of lines at different points on the paper from this one meridional line is a matter of no little difficulty, and such an
operation involves almost certain errors to a greater or less
extent; for whether a rolling parallel ruler, or angle and
straight edge are used (the latter being decidedly the most
accurate) in the operation, errors under such circumstances
cannot be wholly avoided; but with these rectangular and
parallel lines the protractor can be laid down on the part of
the map most convenient for easy plotting, and greater
accuracy can be thus ensured. These lines are also useful in
another way, for if proper attention has been paid to the
preliminary work of the seasoning of the paper; and next,
to the marking of the distances, these lines afford a ready
gauge at all times, by applying the scale to them, to ascertain
whether the Map has been left about in a damp place, and is
unduly expanded with moisture; or whether from having
been in a very dry and warm place, it has contracted below
its normal dimensions.

We have noticed the fact that when a Map has lain open
on the table in a warm sunlight for some time the contraction
has been sufficiently marked to be distinctly appreciable in a
line of 5 inches long; and thus it becomes wise to have regard
to influences likely to affect, in a marked manner, the expan-
sion or contraction of the paper.

By having the plan so divided into squares of 10 acres
each. they also assist the eye in forming a rough estimate of

the quantity of ground exhausted of any particular vein—supposing the ground to have been worked regular and uniform; they also assist the computation, when a more accurate estimate is to be made.

CHAPTER III.

THE MERIDIONAL LINES.

In respect of the "Meridional" lines it has hitherto been generally, or, so far as we know, universally the practice to represent the *magnetic* meridian as the working base line of the plan, from which, at all subsequent times the workings or additions to the plan have been marked off; and this practice has continued in consequence perhaps of, first, a want of due consideration of the actual difference the magnetic variation really makes from year to year in lines of even moderate length; and, secondly, from an absence of a readily and simply adjustable instrument with which to take the angles with the differences corrected.

The western declination of the magnetic meridian is now decreasing at the rate of about $1°$ in 8 years, or about $7\frac{1}{2}$ minutes per annum.

If we may suppose the case of a plan made, say, 8 years ago, with a particular instrument to take the angles, and that in the meantime no addition has been made to some particular part of it; but that now, from a resumption of mining operations or some other cause, an addition is to be made to it from surveys now taken; and if we further suppose a portion of the new workings to be a continuation of a straight engine plane, and that the part to be added measures, say, 850 links; and, further, if the length be laid upon the plan according to a magnetic angle taken *now* with the instrument, we shall find that instead of the present angle corresponding with the last taken 8 years ago, and so showing a continuous straight

line, there will bo a lateral deviation from a straight line to
the extent of 14·83 links to the *left*, looking in the direction
in which the extension has been made. But if it is an
addition to be made after one year instead of eight years,
during which time, at the ordinary rate of decrease, the
deviation will have been 7½ minutes; and if the addition to
the plan be, say, 630 links, the lateral deviation will amount
to 1·36 links for this length ; and when we consider that a
mining plan is frequently being added to for 16 years or
20 years, and often more, the importance of taking the magnetic variation into consideration cannot well be disputed ;
for in 20 years the lateral extreme deviation in a line of 20
chains in length, would amount to no less than 87·2 links ; and
in the simpler case of a 12 *month's* work being added at
one survey at the end of the year, and where the aggregate
of the lines in one direction has been 9 chains, the accumulating error would amount to 1·96 links, or about 16 inches.

These things fairly considered it will readily be seen
what an unsatisfactory base of operations the line of magnetic meridian really is, if treated as an *invariable* line,
while it is, as a matter of fact, *constantly variable* and
changing.

The plan ·of taking the angles with the magnetic needle
is so general, and answers the purpose of a good mining
plan so well, that it is not intended to say anything here
in disparagement of its utility and general applicability;
indeed we do not conceive that there is, at the present time
at least, any other system known, that can be brought into
successful competition with it. It is quite right and proper
to be able at times to dispense with the use of the needle,
when a surveyor is compelled by force of circumstances to
fix the instrument in some locality likely to disarrange the
polarity of the needle, and to interfere with its proper action,
and in such cases it becomes absolutely necessary to dispense
with the use of the needle ; but we conceive it to be in such
cases only that it is found advantageous to dispense with
its use, for we hold that with a thoroughly good instrument
used and handled with care—the needle being set each time

to zero, and the angle being read on the verniers—the use of the needle affords the most correct results that can be attained in mining surveys, and thus the errors which are *instrumental* can be reduced to a minimum.

But as no instruments hitherto in common use have any adjustable zero to alter with the alteration of the magnetic meridian, either once or twice a year, as may be deemed necessary, it has been usual to set the needle to the common zero of the dial-plate, and to work from this line as though it was *invariable*.

Bearing in mind the foregoing facts, we advance the opinion that the meridional lines to be laid down on the plans should represent the *true* meridian, and not the *magnetic* meridian ; and the second series of lines will of course cross these at right angles, and represent the East and West lines upon the plan.

The way in which we propose to provide for the adjustment of the instrument to the True Meridian, and the adoption of that meridian as the working base line instead of the magnetic meridian, will be found described in Chapter VIII. of the present treatise.

CHAPTER IV.

CHAIN, CHAINING, ETC.

THE proper handling of the chain is scarcely of less importance, in mining as in every other surveying, than that of the instrument; and it should not for a moment be supposed that the surveyor can delegate the affairs of the chain altogether to his chainmen.

It is of first importance to have the chain correct in length, and the stretching corrected regularly as found necessary. A correct datum for proving the chain should be at hand, and should frequently be resorted to, to ensure the detection of any extension that may have occurred. It is not sufficient that this should be done at stated periods, it should rather be done whilst bearing in mind the work the chain has been used in; for it will often occur that the chain will have been strained and stretched more with one or two days' operations in some bad place or ground, than in ordinarily good positions in as many months' use; and hence the chain should be tested whenever there is reason for suspecting it necessary, in fact, it should be done often, so that the surveyor may be assured of its accuracy at all times.

This having been done, it should be constantly kept in mind that no amount of care exercised in the taking of the angles with the instrument will compensate for a want of care in taking the lengths of the lines; for a correct angle and incorrect line, or *vice versa*, cannot possibly go to make up a correct plan; hence, however good an assistant the surveyor may have to take the chain-lengths, he should never wholly lose sight of the work himself, but should always give such an amount of attention to it as would ensure a constant diligence in his assistant, to take off the true lengths of the lines; for, especially in underground work, if the chain is very dirty, the difficulty of recognizing the divisions is greatly increased; and, in the like proportion, the vigilance necessary to obtain the correct lengths. As arrows can seldom or never be used underground, the assistant that draws the chain forward has to supply the substitute, in the manner of marking the chain-ends; and as this man will generally be found to be a man picked up in the colliery or mine for the occasion, and with no particular idea of chaining, or any appreciation of the necessity for nicety in marking the lengths, the greater is the danger that it will be thought by him sufficient if he marks somewhere within three inches of the proper place. We have generally found it necessary to give this leading man a short lesson before commencing operations, with the view to ensure his attention and care: and we have found the most accurate way to mark the ends of the chain to be, to cause the leading man, after he has drawn the chain

forward the whole length, and has been put properly into
line, to turn around to face the man at the other or "following" end of the chain; and having strained the chain
properly into line, to bring the *toe* of his right foot to the
handle of the chain,—the handle being held edgewise, and
the foot being placed wholly on the ground; if this be carefully done there can be no error so far. Note is then taken
of any headings, stalls, faults, threads, soft or inferior coals,
commencement of solid rib, or of gob, or of any other thing
on the way requiring to be booked down, and the man at
the "following" end then walks forward and places the toe
of his right foot to the toe of his companion, who then again
proceeds forward another length, and the hindmost handle
of the chain being brought and held to the toe of the foot
the chain is again properly strained, set in line, and again
brought to the right toe of the leading man, and so on to
the end of the line. Where this is regularly and carefully
done, there need be no error of practical moment occurring
in the operation of chaining.

Where arrows are and can be used, the number we have
found to be best is *eleven;* by using eleven there is always
one in the ground, ten being given up to the leading man when
the eleventh has been put into the ground, and it thus becomes
the first in the next ten. This we think to be better than walk-
ing forward when the tenth has been put into the ground, and
then draw it out, so that it may be given up with the nine
others, and the place of the tenth marked with the foot, as in
this way there is the position of the tenth always without its
permanent mark: and so great is the convenience of using
eleven arrows instead of ten, that we conceive the practice
would at once commend itself to any surveyor who should determine to adopt it in preference of using ten only.

It is occasionally found very convenient and advantageous,
when chaining over land on which long coarse grass, rushes, or
other tall plants are growing, to have a strip of coloured cloth or
a small white flag fastened into the handle eye of the arrows,
so that they may be the more readily distinguishable amongst
the tall herbage; and the practice in such cases affords

a means of frequently expediting the work, in avoiding the loss of time which would sometimes be taken up in searching for the arrows, and enables the eye to catch them readily.

CHAPTER V.

FIELD-BOOK.—ENTRIES, POINTS, ETC.

THE method of keeping the surveying book is a matter deserving of more attention than is often given to it, and a looseness or otherwise in this part of the operations indicates, more or less, the kind of attention bestowed on the business generally.

There is a divided practice in the matter of ruling the Field-book, some using two lines within which to write the chain-lengths, &c., and another and less general method is to have one line only,—this line being an ideal representation of the line chained, and the different lengths being written on it. We must give a decided preference to the latter method, as we have found it more generally convenient to have one line only than two ; and in this method the sides of the book are left more clear and open for drawing such lines as may be necessary to illustrate the figures entered down ; and the sketching can be more clearly done than when two lines are used.

In underground surveys, having the book lined ready, and being assured of the correct length of the chain, and having an instrument of good quality, the surveyor having selected the first position to set the dial, will carefully level the same and set the needle to zero after it has become steady ; and before taking the angles the needle should be disturbed by a slight tap on one of the legs of the stand, or, which is quite as well, by lifting the needle off the pivot with the " rest-piece," and lowering it again, so that it may be seen if it will again return to zero, and if it does the reading off the angle may be proceeded with ; if it does not return correctly to zero it should be again slightly disturbed, and if it still varies the first setting was not correct, or something attracts it, and the zero should be moved a little so as to meet that point at which it will permanently set. The limb will now be worked around so that the light may be correctly seen through the sights, and this being done the angle will be read off and entered in the book, with the number of the line set under it : if the limb has two verniers so much the better (in fact any instrument fit for correct surveying is not without two), and these, divided either to 1′ or 3′, the former being much the best, will enable the angle and its opposite reading to be both taken ; and we hold to, and strongly recommend to others, the practice of entering the readings of both verniers in the book, and after they are entered add 180° to the less (mentally), or subtract 180° from the greater reading, and if the sum or difference, as the case may be, makes the other reading as entered in the book, the angles may be considered as correctly entered. Some may think this to be quite unnecessary, but we hold it to be not so, as without it there is no check upon the correct reading of the angle ; and there is no person practising this method but will at times find that an error has been made in entering either one or the other of the readings ; and without the system of check there would be no certainty, and perhaps no probability, of the error being discovered in time to correct it ; but when the two readings are entered, and their difference then computed, there is a strong safeguard against error in this respect.

The two angles of the first two lines having been taken, and the chaining proceeded with, notes are made of anything in the way of headings, faults, or other things occurring in the route, so that the book may be a correct representation of all that will be necessary to be shewn upon the plan : the instrument is then moved forward to beyond the light of the second line, and the operator having gone so far as the back light can be clearly seen, the instrument is again set up, and the operation of setting repeated as before described, and the lines again taken, and so on in succession.

It should be a maxim in working that nothing should be left to memory that can be entered down in the book, for when that is practised the memory will occasionally be overloaded,

and confusion of things will ensue, and error will be the consequence : therefore all things required to be put upon the
plan should be distinctly entered in the book, in such order
that were it to be put aside for seven years, and then taken up
for the purpose of making a Plan of the work described therein, there would be no difficulty in following the particulars
entered, and of making a correct plotting of the same.

An abrupt turn to one side or the other will require a mark
to indicate the direction in which to continue the plotting or
else it may be as natural under certain conditions to think it
should be continued in one direction as the other: for the purpose of indicating the change in direction we recommend the
making of a mark thus <-   , under the figures (for instance ;___⍺___) giving the number of the line *commencing
at the turn.* The horizontal line is simply that line which is
drawn between the figures expressing the *total length* of the
preceding, and the *number* of the succeeding line ; and the
mark either < or > indicates,—the former a turn to the *right*
and the latter a turn to the *left*, it being in the latter case
placed on *the right* of the horizontal line, and in the former
case on *the left*. This mark must not be confounded with the
arrow or else it will mislead : thus, the mark ———— , here
recommended, indicating a turn to the right, and ——· ·—⁄
indicating a turn to the left, may, if made into arrows thus
·⧼———— , or ————⧽ , be made to indicate the opposite ;
but no difficulty on this head will arise when the distinction in
the marks is remembered ;—the idea being that the point
joining the two short lines of the mark <, or >, represents the
position of the instrument in reference to the two adjoining
lines, and the short lines meeting in that point shew the
relative directions of the lines so meeting, of which they are
the representatives.

Note must also be taken of the "rise" or "pitch" of the
measures in which the chaining is being done, so as to allow
for the same, and to eliminate the true horizontal measure.

An example is given on Plate 4 of a book made upon
these principles, with some descriptive entries, and a plotting
of the same for the guidance of students.

It will sometimes occur that a survey is required to be made so as to give "points" to headings being driven towards each other, from distant parts of the workings, and it may be of great importance that they should meet well in line, not only to save the cost of driving about (perhaps in hard ground) to get "loose," but to ensure the working being subsequently usable for a heading, engine plane, or some other important work ; or it may be to find the relative position which a proposed pit, either for air or otherwise, occupies, or where it will come down when sunk from some overlying measures or veins; or it may be to ascertain the proper position and direction for a drivage into a pit in some measures through which the pit has passed, for drainage or for ventilation : in these and similar cases it will be advisable to make an entire survey around, from one face of the work to the other, or so far as it is practicable to go ; and in such surveys it will seldom be necessary to make notes of anything particular on the way, except for the purposes of a check at a few points. These lines will require to be plotted with great care, so as to be able to determine the "point" that will have to be given to the drivages, so that they may meet well. It is in such cases and similar to these, that the method of working out the line of survey by the Traverse Tables is so important, and of such utility; for if the survey has been correctly made, and the lines worked out by the Tables, the *terminal* positions can be determined with much greater accuracy by this combined operation than by plotting. We would not indeed recommend that the plotting should not be done; but where the survey was both worked out by the Tables and plotted, and some little difference resulted in the two methods (which may certainly be looked for), we would unhesitatingly give the preference for accuracy to the result obtained by the Tables. If the difference should be very considerable there may be some error either in the plotting, or in the taking out the figures of the Tables, or in not placing them in their proper columns, and a search should be made to discover it, so that the greater confidence may be given to the result obtained. But we have had a difference between the plotted and the computed plan (or the terminal points of such plan) of considerable length, of upwards of 30 links ; and having failed to discover any error in either the operation of plotting, or working by the Tables, we have decided in favour of the result obtained by using the Tables, and have found the center of the sinking (the matter in hand in that case) to come down within two feet of the mark made for it, although the *route* of the survey lay through very crooked and uneven ground, and through a pit from one measure to the other.   This is not given as an unusually good result, for better could be added, but only as a proof that had the plotted survey been acted upon—although it was laid down with the greatest care—instead of the result obtained by the Tables, the error in this case would have been considerable ; but as a rule the difference would be less, except in long and difficult surveys when it would perhaps be even more.   And it is only what would justly be expected in such cases, if we consider the method of computation to be more correct than plotting could possibly be ; for there are no thick compass points, no little accumulating errors in the setting the lines from point to point, the thickness of pencil lines etc., but the rectangular co-ordinates of each line are joined by absolute mathematical points.

If the opposite drivages are to be a heading or drift in hard ground, instead of following a vein of coal or other mineral, it will then be necessary to "level" the headings around from one point of driving to the other, so that the difference of level may be also known, and the "pitch" the new drivages are to be set to, worked out.   The miners should then be supplied with teebobs, inclined to the proper angle at the foot, with which they should set the plates or rails of their roads ; and as this is only an approximation to the true pitch, the surveyor should take his Spirit-level and try the accuracy of the floors occasionally, as well as examine the "points" by the dial, to ensure against error in either pitch or points.

Great caution should be exercised in making use of timber for holding the points ; it will often happen that in consequence of one side of the heading being a little "tighter" than the other, at the place where timber has been selected to hold the point marks, the squeeze of the ground will cause a slight (and sometimes a considerable) lateral movement in one or both of the colar pieces so selected; and possibly they may have motion imparted to them in opposite directions, as is frequently seen; and if such should occur to two colars used for carrying the "point," or even to one of them, the consequence may be serious if not detected in time to rectify the error. Therefore the solid ground should, if possible, carry the points, even if it should be necessary to stand timber for the sole purpose of preventing the roof scaling off, or for supporting the ground from falling while the drivages are in operation, or until the "points" can be moved to more secure ground.

In moving "points" forward great care should be practised, so as to ensure the continuance of the work in the desired alignment. If it is a drivage going to meet another coming from the other side, or a single drivage going to some particular position in some workings ahead of the new heading, the instrument should be used to move the "points" forward when it becomes necessary: and the dial being placed under the first of the former point-marks, others can be set up near the face of the workings for further guidance; or, which is better, a mark should be put at about every 30 yards along the heading or drift, and from these the men should try their work. It is much more sure to drive with points 30 yards apart, if they are carefully put in position, than with those at 10 feet apart; and in the former arrangement it is only necessary to shift one mark at one time, and they are always within easy reach of the workmen.

When two marks are to be shifted forward to the "face" without the application of the dial, the accuracy of the new marks should be carefully tested; and to do this proceed by setting up the new points near the face by plumbing marks to the roof, from lights put in position by looking through the original mark lines: then, first, have a light placed in the "face," by using the *original* mark lines; next, by using the

new lines see if they and that in the "face" still coincide
(which they should do); next, take *one* of the old lines and
one of the new, and again see if they coincide with that in the
"face"; if they do this last operation will be a proof that the
new lines will correctly continue the alignment given with
the old lines, and the marks will have been correctly shifted.

As to the kind of appliance that is best for holding the
light, we may remark that we have found it the better way
to have no appliance at all for suspending the light, but
simply to weight the cord, soundly fastened to the roof,
sufficiently to draw it out straight and free from kinks. If
then the light be held right in front of it, if marks 30 yards
distant be used; or by the side and at two inches from it, if
the marks are about 10 feet apart,—for at the latter distance
the string itself will be plainly visible with the candle two
inches off,—the arrangement will admit of very easy and correct working.

If however safety lamps are exclusively used, and open
lights are inadmissible, it is then better to suspend the
safety lamps from the lines by their handles, and use them
in that way to move the mark forward; but in all cases
more care will be necessary in driving with safety lamps, than
with free lights, as it is more difficult in that case both to
"try" and also to keep the point.

The benefit which should and would result to the employer
or owner, as well as the satisfaction felt by the surveyor at
seeing the work "meet" correctly, will more than compensate
for what some may be disposed to look upon, but which we do
not, as excessive nicety in the matter of "points" and "pitches";
and not only in these matters, but in every other matter connected with correct surveying, it is the attention to the small
details, and the care with which they are executed, that
make the difference between a good plan and a comparatively
worthless one : for even mining surveying (the roughest-done
work of any perhaps in the way of surveying) consists in
something more than in being able to tell, within an half
degree, the division on the dial to which the needle points,
though this is often looked upon as all that is necessary to

enable a person to survey a colliery; but the "snakelike" contortions and windings often to be met with in collieries and mines, as the consequence of not being able to get opposite drivings *loose* in the way they were intended to meet, testify to the rough off-hand methods too often practised, as well as to the expensiveness of such a class of surveying.

CHAPTER VI.

TRAVERSING.

HAVING passed in review the field-book, setting of the dial, reading off the angles, taking the notes, and other things necessary to supply the data and particulars requisite to the making of a correct plan of what will be met with in the mine or colliery, we will next proceed to show what entries are requisite and essential, in order that the Traverse Tables may be used in the computation of a survey.

In the first place we will refer to the accompanying diagram, No. 1, and would recommend a very careful perusal of the whole of the lines comprising the same, and especially of the red lines, and the directions in which the "arrows' point. Let it be supposed that the positions A and B represent two pits or other objects, and that the black lines represent the lines chained in passing, in the operation of surveying, from A to B; note also the direction of the meridional line with its transverse axis, and the point that represents the north. The black lines 1, 2, 3 will be observed to go nearly eastward; lines 4, 5, and 6 nearly northward; 7, and 8, have a northern departure from the west, and 9 goes to the south of west; 10, 11, and 12 then go nearly to the north; 13 goes nearly east, and 14 nearly southward. Now it is possible to define the relative positions of A and B without plotting the lines at all, by the use of the Tables, and the simple operations of addition and subtraction.

c

In conjunction with each black line representing the chained lengths, will be noticed two red lines, each containing an arrow, and forming with the black line a right-angled triangle. All the *red* lines parallel with the *meridian* line of the diagram are represented by figures in the columns of the Tables headed Lat. (a contraction of " Latitude ") and Per. (a contraction of " Perpendicular"), and their values will be taken out from that column; and all red lines parallel with the *transverse* axis of the meridian, being parallel with the equatorial line, are represented by figures in the columns headed Dep. (a contraction of "Departure"), and Base, and their values will be obtained from those columns. Next observe that some of the arrows in the red lines parallel with the meridian point to the North, and some to the South, accordingly as the line to which they stand points northward or southward of the east and west line that may be conceived to pass through the zero or starting-point of the line in question, of the diagram: some of the arrows parallel with the equatorial line point eastward, and some point westward, accordingly as the line to which they stand points eastward, or westward of the corresponding ideal meridian line. Now if we first take the equivalents of the red lines parallel with the meridian, and separate those pointing to the north from those pointing to the south, or, which is the same thing, place them in separate columns, and take the total of each column, we shall see in which direction the working has progressed by observing which column has the higher total; and if the less be subtracted from the greater, the difference or resultant will be the actual distance between the two terminal points, *taken in a line parallel with the meridian.* Also it will be similarly observed that of the lines parallel with the equator —some pointing to the east, and some to the west—if we separate the equivalents of these into distinct columns as we did the other, and take the totals of each, and their difference, we shall find what is the resultant sum of the progress in a line parallel with the equator, and whether it has been to the east or to the west.

These two differences will now give the data with which

to define the relative positions of *A* and *B*. Making a mark to represent *A*, and from this mark drawing a straight line of sufficient length to enable us to prick off on it by some scale (any may be chosen) the first found difference, make a fine and distinct mark; then with a set square or other accurate instrument, draw carefully at right angles from this point a line, either to the right or to the left as required by the *second resultant*, and on this line mark off with the same scale as before the second difference; this distance from the first line will be the position of *B*. The former of these lines will be a north and south line, and the latter will be an east and west line; thus the required relative positions are determined, and the lines with which it is done form the perpendicular and base of a right-angled triangle.

The thorough digestion of the remarks and explanations here given, in reference to the diagram and its several lines, will make the apprehension of what follows comparatively easy; and it is essential that it should be thoroughly understood, before proceeding to attempt to work out any series of lines by the Tables.

Traversing Entries in Field Book. The next thing will be to shew how the angle is to be entered in the book, so that there may be no difficulty in easily determining the proper reading.

The plate forming the *bottom* of a dial is usually divided and figured in each quadrant distinct, from $1°$ to $90°$; two quadrants commencing to reckon from the *N.* point in the box, and two commencing at the *S.* point; two finishing at the *W.* point, and two finishing at *E.* This division of the circle into quadrants shews at once, in their respective initial and terminal points of division, *the points from which*, and the *direction in which* each quadrant must be read, in order to the work-ings being computed by the Traverse Tables.

The first quadrant therefore, embracing the horizon from north to east, will be read *from* north *to* east; the second quadrant, embracing the horizon from east to south, will be read *from* south *to* east; the third quadrant, embracing the horizon from south to west, will be read *from* south *to* west;

and the fourth quadrant, *from* north *to* west. This must be the *invariable practice* for traverse working, and where the readings cannot be obtained from the verniers, as they cannot be in more than the first quadrant, the angles must be carefully computed: and for this purpose the readings should be severally entered in the surveying-book as before described, as though it was intended to make use of them in plotting a survey as ordinarily: then, the traversing angle in the first quadrant will be the ordinary reading of the instrument; the traversing angle in the second quadrant will be the difference between the ordinary reading and 180°, thus 100° 42′ in the ordinary reading will be 79″ 18′ for traversing, and 146° 27′ in the ordinary reading will be 33° 33′ for working by the Tables; the traversing angle in the third quadrant will be the ordinary reading of the instrument, *less* 180°, thus 212° 30′ in the ordinary reading becomes 32° 30′ for traversing, and 254° 12′ will become 74° 12′ for working by the Tables; and the traversing angle in the fourth quadrant will be the difference between the ordinary reading and 360°, thus 293° 45′ will become 66° 15′ for traversing, and 357° 21′ will become 2° 39′ for working with the Tables; and so on of all other angles, in whichever quadrant they may be situate. There is here no great difficulty in comprehending this part of the matter clearly, close diligence and thought being only necessary.

The next thing to be mastered is the *description* to be given in the traversing to the *co-ordinates* of each line: by the description to be given them is to be understood the entering with each angle distinctively whether its co-ordinates are a northing and a westing, or a southing and a westing; a southing and an easting, or a northing and an easting. Thus the angle in the first quadrant (remembering the order given in the instructions above in reference to the angles) will have a northing and an easting for its co-ordinates; the angle in the second quadrant will have a southing and an easting for its co-ordinates; the angle in the third quadrant will have a southing and a westing for its co-ordinates; and the angle in the fourth quadrant will have a northing and a westing for its co-ordinates; and these rules hold good for every angle in the respective quadrants in succession. Of course if the line should be *exactly* due north, or due south, it will neither have an eastern nor a western co-ordinate, but it will be a northing or a southing entire, without reference to the Tables at all; and in the case of a due east or a due west line, there would be no northing or southing co-ordinate, but the line would be an entire easting or westing as the case may be, without any reference to the Tables, the lengths of the lines to which they refer being entered in full.

These explanations and instructions embrace the whole that need be said to enable any person of ordinary intelligence, and who will try, to understand the theory and practice of working with the Traverse Tables; but the instructions here given should be thoroughly mastered, so that the Tables may be correctly worked, and errors avoided in the operation.

The entry of the particulars for traversing should be made upon the same sheet or leaf of the field-book with the ordinary entries, and will simply consist of the number of the line, the traversing angle, the initials of its descriptive co-ordinates, and the total length; and these can be placed in the side of the book, as in the example we give herewith of a field-book, with both entries placed in juxtaposition.

CHAPTER VII.

THE TRAVERSE TABLES: THEIR CONSTRUCTION, AND APPLICATION.

THESE Tables were constructed by the author some years ago for use in his professional duties, and they are based on the same principles as the Traverse Tables for seamen's practice as given in " Hutton's Mathematical Tables." It will however be found, on reference to that book, that the Tables there given are worked out for degrees and quarter-points only, and to radii from 1 to 10; and as a Table not less minute than this is inapplicable to surveying we determine to proceed to the construction of tables to every 3 minutes of the quadrant, and to 10 different radii of hundreds, commencing with 100, and ending with 1000. It will be readily seen that thus constructed the Tables are applicable to any base of measurement, either English or Foreign, that may be most convenient; it may be a link, foot, yard, furlong, mile, metre, kilometre, or league, in fact any measure desirable or necessary; but it is of course requisite that the measure of the *angle* should be taken according to the *English* system of 90 degrees in the quadrant, and 60 minutes in the degree.

The figures in the Table, and the positions of the decimal points, are applicable to the lengths represented at the top and bottom of each pair of columns, and marked "Dist."; these are each hundreds. But the computations are perfectly correct and applicable if the "Dist." is multiplied by 10, or considered as thousands; and when this is done it is only requisite to shift the decimal point as in the Tables, or conceive it to be shifted, one figure to the *right*, when the result will be obtained true to one place of decimals. If the "Dist." be *divided* by 10, the figures as in the Tables will in like manner be divided by 10 by shifting the decimal point one figure, or conceiving it to be shifted, to the *left;* and if divided by 100 (say by counting 600 as 6) the figures as in the Tables will be in like manner divided by 100 by shifting the decimal point *two places* to the left: hence the Tables are applicable to the computation of the co-ordinates of a line of any measurement or length whatever, from 1 link to 100,000 links, or from 1 metre to 100,000 metres, or kilometres, or miles. The Tables being computed to every 3 minutes of the quadrant will in general be sufficiently minute for all classes of surveying; but in the case of very long lines, if it should be found or thought necessary to obtain the co-ordinates to a single minute, it can be done to very near the truth (even when the decimal point is moved one figure to the right) by interpolation, taking either ⅓ or ⅔ of the difference as may be required, between those two lines of figures within which the angle to be computed lies (in doing this remember that the co-ordinates of each line *increase* in opposite directions, one increasing down the column, and the other up the column).

The degrees from 1° to 45° will be found along the top, and the minutes down the left side of the tables; and from 45° to 90° the degrees will be found along the bottom, and the minutes up the right side of the Tables, as in Trigonometrical Tables generally. If the angle therefore is less than 45° it will be found at the top and left side of the Tables, and the proper order for tabulating the co-ordinates of the angle will likewise be found at the top; if the angle is greater than 45° it will be found at the bottom and right side, and the order for tabulating the co-ordinates at the bottom likewise.

Thus, if we wanted to find the co-ordinates of a line 600 links in length, and whose ∠ is 33° 42', we have on the page for 33° and on the line for 42' the quantities sought; viz—Per. = 499·17., and Base = 332·90; also, if we want the co-ordinates of ∠ 56° 18' (or the comp. of 33° 42'), and a line of the same length as before, we have Per. = 332·90, and Base = 499·17;—being the same figures as by the first operation, *but with their relative positions reversed.*

If the lines had been each of them 666 links, instead of 600, we should have had in the first case, viz. ∠ 33° 42' and Dist. 666, as follows.

Ex. I.

	Per.		Base
600 =	499·17	and	332·90
60 =	49·92	„	33·29
6 =	4·99	„	3·33
Totals.	Per. = 554·08		Base = 369·52

And in the second case we should have ∠ 56° 18' and Dist. 666.

	Per.		Base
Ex. 2.	600 = 332˙90	and	499˙17
	60 = 33˙29	„	49˙92
	6 = 3˙33	„	4˙99
Totals =	369˙52	and	554˙08

Again, suppose the case of ∠ 20° 39′, and Dist. 745 ; then look along the Tables until you find 20° and down the left side to 39′, and on that line take out the figures for 700, 40, and 5 respectively, thus :—

Ex. 3.   Angle 20° 39′, and Dist. 745.

	Per.		Base
	700 = 655˙02	and	246˙86
	40 = 37˙43	„	14˙11
	5 = 4˙68	„	1˙76
Totals =	697.13	aud	262˙73

Take one more example. ∠ 76° 45′, Dist. 1789 links. In this case the degrees will be found along the bottom, and the minutes on the right side thus :

	Per.		Baso
Ex. 4.	1000 = 229˙20	and	973˙38
	700 = 160˙44	„	681˙36
	80 = 18˙34	„	77˙87
	9 = 2˙06	„	8˙76
Totals =	410˙04	and	1741˙37

The foregoing examples will clearly indicate the methods of taking out the co-ordinates for traverse surveying.

It will be perhaps advisable to draw attention to the last example, and to point out that in consequence of the Tables having computations for 1000, no shifting of the decimal point is called for in this case in reference to that number ; but for every 1000 above the first there will be a shifting of the decimal point;—for instance, the co-ordinates for 2000 will be found in the columns for 200; with the decimal point placed one figure to the right, and so on for the others as each case may require.

It should also be the invariable practice to place the figures coming under the designation of· "Per." *first* in position, as they are above, and those coming under the designation of "Base" *last*. It should also be remembered to consider the co-ordinate designated "Per.", as that ordinate *parallel with the meridian;* and the co-ordinate designated "Base" as that *parallel with the equator:* by following these instructions there will be no confusion of the co-ordinates, nor any misconception of the true positions for the termini to be described from the results obtained.

We will now adduce proof of the foregoing examples by working out corresponding results by the ordinary rules applicable to the computation of right-angled trigonometry, as in the following.

RULE: *Add the logarithm of the given side to the sine of the angle opposite to the side required, and from the sum subtract the sine of the angle opposite to the given side: the remainder will be the logarithm of the side required.*

Bear in mind that the sine of the *angle of the traverse* is the co-ordinate "Base"; and the sine of its complement, the co-ordinate "Per." (or "Perpendicular").

The sine of $90°$ being equal radius—the angles treated being all right angles—we have

One proof of Ex. 1. (giving the "Base")

As Rad............................ $= 10\cdot0000000$

: log. of 666 .... $= 2\cdot8234742$

:: sine ∠ $33° 42'$................ $= 9\cdot7441712$

: log. of 369·25 .... $= 2\cdot5676454$

Also one proof of Ex. 2. (giving the "Base").

As Rad........................ ... $= 10\cdot0000000$

: log. of 666 .... $= 2\cdot8234742$

:: sine ∠ $56° 18'$.............. $= 9\cdot9200994$

: log. of 554·08............. . $= 2\cdot7435736$

Proofs of Ex. 3.

$$
\begin{aligned}
\text{As Rad} &= 10\text{·}0000000 \\
: \log. \text{ of } 745 &= 2\text{·}8721563 \\
:: \text{sine } \angle 20^0\ 39' &= 9\text{·}5473542 \\
: \log. \text{ of } 262\text{·}73 &= 2\text{·}4195105
\end{aligned}
$$

and

$$
\begin{aligned}
\text{As Rad} &= 10\text{·}0000000 \\
: \log. \text{ of } 745 &= 2\text{·}8721563 \\
:: \text{sine } \angle 69^0\ 21' &= 9\text{·}9711608 \\
: \log. \text{ of } 697\text{·}13 &= 2\text{·}8433171
\end{aligned}
$$

Proofs of Ex. 4.

$$
\begin{aligned}
\text{As Rad} &= 10\text{·}0000000 \\
: \log. \text{ of } 1789 &= 3\text{·}2526103 \\
:: \text{sine } \angle 76^0\ 45' &= 9\text{·}9882821 \\
: \log. \text{ of } 1741\text{·}37 &= 3\text{·}2408924
\end{aligned}
$$

and

$$
\begin{aligned}
\text{As Rad} &= 10\text{·}0000000 \\
: \log. \text{ of } 1789 &= 3\text{·}2526103 \\
:: \text{sine } \angle 13^0\ 15' &= 9\text{·}3602154 \\
: \log \text{ of } 410\text{·}04 &= 2\text{·}6128257
\end{aligned}
$$

These proofs shew the construction of the Tables, as well as supplying conclusive evidence of their accuracy, and of the accuracy of the method employed in the operation of extracting the co-ordinates for the different angles, and length of the lines.

There is nothing new in the proposal to apply these principles to surveying. They were, so long ago as A.D. 1635, applied by a Mr Norwood to lines and angles taken with the circumferentor over the public road from York to London, and the difference of latitude of those two cities determined by him, very nearly, from the traverses; and proposals have been made to apply the principles to the planning of large estates; but, so far as the author knows, there have been no other Tables so far issued, so well calculated as these are to render the application of these principles to surveying both easy and general,—computed as these are to a degree of minuteness sufficient for all practical purposes of correct surveying, and their application duly explained in the accompanying letter-press.

There is no reason why these principles should not be applied extensively to mining surveying; for in large and overlying mining operations there are cases continually arising where it is required to sink from one vein of mineral to another, for ventilation or some other purpose; also to drive from one part of the workings in some mine or colliery across the measures to another; and also when driving in both directions it is highly important that the points given to the drivages should be correct; and sometimes it is required to drift at a more or less steep pitch from one measure to another; and in such operations, when the work proceeds in both directions at the same time, the question of the cost, as well as the matter of convenience, depends greatly upon good surveying and correct points; and it is such cases as these that really put the skill of the surveyor to the test, and they are also those cases in which the application of the principle of traversing is found to yield the most satisfactory results. But cases arise also where it is desirable to ascertain if the boundary has been reached, at some point very remote from the Pit, or level mouth; and sometimes it will require a series of lines of somewhere between 60 and 100 (perhaps more) to reach the supposed position of the boundary: then the survey should be taken with great care, and the application of Traversing should be adopted in such a case, as being much more conducive to accuracy than plotting can be from the same data.

It will likewise occur frequently that in a long underground survey there will occasionally be a prominent turn in the general bearing of the workings; perhaps the general bearing so far has been nearly westward, and at a particular part the direction alters, and the bearings are somewhere to northward. It is very desirable that the point where the change of direction occurs should be very carefully marked down from the traverses first, and also the terminal point ; the plotting should then be proceeded with, and the plan so made should be brought as near as possible at those corresponding parts to the points marked from the traversing computation ; and where a little difference is found to have accrued, preference should be given to the traverse points (always bearing in mind that care should be observed to ensure accuracy in taking out the co-ordinates).

If it is suggested that this double operation is unnecessary, and that plotting is sufficient without traversing, we can confidently assert that bad and loose surveying is exceedingly expensive ; and that there is no system so conducive to true economy as accuracy, and this cannot be attained without due measures being adopted to ensure it.

The chief utility of the Traverse Tables in surface surveying is the facilities they afford for reducing the principal lines across an estate to one pair of rectangular co-ordinates, and thus increase accuracy.

Although it is still the practice with many surveyors, in making a survey of a property, to make a circuit of the boundary by taking a great number of short lines according to the bends, or outline of the limits of the property,—making these short lines the ground-work of the survey from which to construct the map,—and especially where there is but little more than the boundary to lay down; but it is about the worst system that can be followed, and often results in the production of a Plan not worth the labour expended upon it.

The salient angles in the boundary of the property to be surveyed should first be selected by the surveyor, in sufficient number to enable him to devise such a system of main base-lines and triangles across the property as shall enable him to fix all those points, as well as a sufficient number of objects within the boundary on his Plan, with absolute certainty, by Traversing, or Trigonometry, or both combined, as circumstances require : the intervening lengths of the boundary should then be surveyed by a combination of straight lines and triangles, according to the conformation of the boundary line ; and in all cases running the straight lines *where practicable* from one salient point to another, and filling in where necessary with diverging lines to meet bends, and which divergence may be measured as a check with a sextant. In this way, if errors have been committed at any point they are almost sure of being detected, and their locality known; and if they arise between some two of the points selected, they cannot affect the accuracy of the next section; and all the internal objects will be easily taken, in connection with the different internal stations of the triangulation.

As scarcely any two estates are of similar outline, or are affected by similar conformations, so the same system of lines will seldom be found perfectly applicable to any two surveys, and hence the judgment and experience of the surveyor must in each case be relied on to guide him, in the selection of that system which will involve the least amount of labour combined with the greatest amount of attainable accuracy.

Where the estate is tolerably level the selection of these lines, and their measurement, do not present much difficulty; but where the country is mountainous, and steep escarpments have to be crossed and measured it is different, and careful means have to be then adopted to ensure accuracy. Where it appears to be indispensably necessary to have a line, either as a principal, or as a check-line, up a steep and rugged rise, it will generally be found more accurate and preferable to obtain the true length of the line by the application of trigonometry; and which will probably be best accomplished by selecting two stations in the plain, one of which shall be the end of the line to be measured (or even an intermediate point in it, if it is thought to answer the purpose better than the end), and between which and the other station in the plain the ground must be favourable for *correct chaining,*—each visible from the other, and from another station on the top of the hill, which may be the other end of, or an intermediate position in, the line to be measured, and to be selected on favourable ground, from which correct chaining may be continued. The three angles of this triangle should be correctly taken with an accurate instrument, and their sum should equal 180°; and the line in the plain to be measured should be gone over carefully twice, so that the first result may be checked. Then, by applying the rule of trigonometry previously given in this chapter, the correct horizontal length of the ascending line can be easily determined; and this length should be substituted for the length found by chaining, as being more correct, (even though the horizontal measure may have been procured as near as practicable), and therefore entitled to the preference. If this is a line forming one of a series, for the purpose of determining the positions of the *salient points of the boundary only*, the *chaining* of it will be unnecessary.

When it is thus determined to have a few triangles over the property, to determine the position of the salient points, from which to form the ground-work of the Plan, one base-line should be measured on the most favourable ground obtainable near one side of the property; this should be the "base of computation," on which all the calculations of the Triangulation shall be made; and on the opposite side of the

TRAVERSING COMPUTATION. No. 1.

No. of the line.	Quadrant where found.	Traverse Angle.	Designation of Co-ordinates.	Length in Links.	Perpendicular.		Base.		Remarks.
		° ′			North-ing.	South-ing.	East-ing.	West-ing.	
1	1st	78.26	N. E.	208	41.71	...	203.78	...	
2	2nd	76.42	S. E.	354	...	81.43	344.50	...	
3	1st	3.46	N. E.	246	245.47	...	16.15	...	
4	4th	5.34	N. W.	208	207.02	...	...	20.17	
5	do.	11.48	N. W.	349	341.62	...	...	71.37	
6	do.	15.00	N. W.	540	521.60	...	...	139.76	
7	do.	81. 9	N. W.	240	36.92	...	...	237.14	
8	3rd	85.30	S. W.	320	...	25.10	...	319.01	
9	4th	61.33	N. W.	528	251.53	...	...	464.22	
10	1st	3.00	N. E.	340	339.53	...	17.79	...	

Totals are = 1985.40 | 106.53 | 582.22 | 1251.67

Subtract   106.53   582.22

Nett results 1878.87 = Northing.   669.45 = Westing.

property the best line possible for correct measurement should be chained, as a check upon the trigonometrical computation. This will be called the "base of verification," and if the angles

---

The following will exhibit the operation of taking out the above co-ordinates in succession.

(1) 78'·26' and 208		
Dist.	Per.	Base.
200 =	40·11 −	195·94
8 =	1·60 −	7·84
208 =	41·71 −	203·78

(2) 76°·42' and 354		
Dist.	Per.	Base.
300 =	69·01 −	291·95
50 =	11·50 −	48·66
4 =	·92 −	3·89
354 =	81·43 −	344·50

(3) 3°·46' and 246		
Dist.	Per.	Base.
200 =	199·57 −	13·13
40 =	39·91 −	2·63
6 =	5·99 −·	·39
246 =	245·47 −	16·15

(4) 5°·34' and 208		
Dist.	Per.	Base.
200 =	199·06 −	19·40
8 =	7·96 −	·77
208 =	207·02 −	20·17

(5) 11°·48' and 349		
Dist.	Per.	Base.
300 =	293·66 −	61·35
40 =	39·15 −	8·18
9 =	8·81 −	1·84
349 =	341·62 −	71·37

(6) 15° and 540		
Dist.	Per.	Base.
500 =	482·96 −	129·41
40 =	38·64 −	10·35
540 =	521·60 −	139·76

(7) 81°·9' and 240		
Dist.	Per.	Base.
200 =	30·77 −	197·62
40 =	6·15 −	39·52
240 =	36·92 −	237·14

(8) 85°·30' and 320		
Dist.	Per.	Base.
300 =	23·53 −	299·07
20 =	1·57 −	19·94
320 =	25·10 −	319·01

(9) 61°·33' and 528		
Dist.	Per.	Base.
500 =	238·19 −	439·61
20 =	9·53 −	17·58
8 =	3·81 −	7·03
528 =	251·53 −	464·22

(10) 3° and 340		
Dist.	Per.	Base.
300 =	299·59 −	15·70
40 =	39·94 −	2·09
340 =	339·53 −	17·79

nave been taken and the sides calculated, each correctly, this last measured line should not differ from the computed length more than from 2 to 3 links. This satisfactory result being obtained (and with instruments of the best construction for taking the angles a nearer agreement is quite possible, and generally considered essential), the points so determined will form an excellent ground-work for the making of a thoroughly reliable map of the property; and the labour expended in obtaining it will be well repaid, by the confidence with which the Plan may subsequently be referred to at all times; and especially should any question as to infringement of boundary, in the working of the minerals of the property, or a committal of trespass from an adjoining property, at any time arise.

As shewing the application of Traversing to a series of underground lines of a survey, we will now proceed to take the first 10 lines of the preceding example of a Field-book, in which *A* (see the plotted Plan of same), the starting point, may be supposed to represent a Pit or Level Mouth; and *B* to be the "face," and end of the 10th line. *C* may also be considered as the face of a branch part of the same workings; and if we suppose it to be required to drive from *C* to *B*, by having a point given at *C*, we shall be able by the Tables to discover the true distance from *C* to *B*, and the point to be given at *C* to drive into the "face" at *B*.

The nett results of the foregoing traversing operations is a Northing of 1878 87 links, and a Westing of 669·45 links; and if the process as before directed be performed, the terminal point of the (10) line can be exactly described from the above co-ordinates as here found and expressed.

We may now proceed from the two lines here found of the triangle, with the included right angle, to determine the other line and the two other angles, and thus complete the triangle; the rule for which is the following: Having two sides and the included angle, .

RULE:—*When the angle included is a right angle, add the radius to the logarithm of the less side, and from the sum subtract the logarithm of the greater side, or add its arith.*

complement: the remainder, or sum, will be the tangent of the angle opposed to the less side.

Then, taking the foregoing sides of the triangle, we have

Rad. + Log. of $\quad$ 669·45 = 12 8257181

Arith. Comp. Log. 1878·87 = 6·7261032

$$\overline{\qquad 9·5518213} = \text{Tan. } 19^\circ 37' \text{ nearly.}$$

Therefore $90^\circ - 19^\circ 37' = 70^\circ 23'$ is = the third angle of the triangle.

By applying Euclid I, Prop. 47, we shall ascertain the length of the third side: using logarithms, we have

Log. 1878·87 = 3·2738968

$$\underset{\overline{\qquad\qquad}}{2}$$

$$6·5477936 = 3530153·3 = 1878·87^2.$$

Log. 669·45 = 2·8257181

$$\underset{\overline{\qquad\qquad}}{2}$$

$$5·6514362 = 448163·2 = 669·45^2.$$

And the sum of the squares are thus = 3978316·5

Then, Log. 3978316·5 = 6·5996993

$$\div \text{ by } 2 = 3·2998496 = 1994·57 = \sqrt{3978316·5}.$$

= the side *opposite the right angle.*

We may prove this result by finding the number 1994·57 from the Tables: taking 19° 36′ as being the next nearest number in the minute column we have,

	Per.	Base.	Numbers.
∠ 19° 36′, and the coordinates	1878·87	669·45	
in column of thousands take	942·06	335·45	= 1000
remainder	936·81	334·00	
„ „ of 900 take	847·85	301·91	= 900
remainder	88·96	32·09	
„ „ of 900 take	84·78	30·19	= 90
remainder	4·18	1·90	
„ „ of 400 take	3·76	1·34	= 4
„ „ of 500 approximately	·42	·56	= ·5
			1994·5

These numbers from the Tables being successively subtracted from the preceding remainder, exhaust the co-ordinates; and, adding up the successive corresponding numbers, we have 1994·5.

This last operation proves the correctness of the mode of working, and also the finding the side by Euclid i, Prop. 47; for although the last figure ·5 is not found strictly correct, yet, as the tangent of the angle lay between 19° 36′ and 19° 37′ (nearer the latter than the former), and the co-ordinates here taken out stand on the line for 19° 36′, the agreement is so near that for all practical purposes the difference may be disregarded.

It may sometimes be necessary to find the third side, and the two other angles, so that the position of the third line may be correctly laid out on the surface from one of the termini (which may perhaps be the center of a Pit or the Mouth of a Level) to the other, and the angle set out for finding the position of the other terminal point corresponding with that underground, either by chaining along the line laid out to the angle given, or by the application of trigonometry, or a combination of both: this method is much more correct than that of attempting to repeat on the surface all the lines taken underground from one terminal point to the other. Where an important work, such as the sinking of a Pit or some other matter is to be proceeded with, upon the data arrived at by such methods as these, care and precision, with the aid of good instruments, should yield perfectly satisfactory results.

It will here be proper to remark that the lesser angle, whose tangent is found by the foregoing trigonometrical computation, is not *necessarily* the "traversing" angle, by which to ascertain the length of the third side, as in this example; for "the greater side of a triangle is subtended by the greater angle," *but the traversing angle is always subtended by the co-ordinate* "Base": this will be either the greater or lesser co-ordinate according to the nature of the lines operated on In this example the co-ordinate "Base" is the lesser, and therefore subtended by the lesser angle; and as this trigonometrical operation is to find the lesser angle, it in *this case* finds the *traversing angle*. But supposing 1878·87 to have been the co-ordinate "Base", and 669·45 to have been the co-ordinate "Per.", then in that case 70° 23' would have been

TRAVERSING COMPUTATION. No. 2.

No. of the line.	Quad-rant where found.	Tra-verse Angle.	Desig-nation of Co-ordi-nates.		Length in Links.	Latitude.		Departure.		Remarks.
		° '				North-ing.	South-ing.	East-ing.	West-ing.	
1	1st	78·26	N.	E.	208	41·71	...	203·78	...	
2	2nd	76·42	S.	E.	78	...	17·94	75·90	...	
11	4th	56·42	N.	W.	246	135·05	...	...	205·60	
12	4th	64·50	N.	W.	332	141·18	...	...	300·48	
13	3rd	84·36	S.	W.	450	...	42·34	...	448·00	
14	3rd	74·30	S.	W.	320	...	85·51	...	308·36	
15	Par'.	90·00	...	W.	248	...	..	...	248·00	
16	4th	10·51	N.	W.	675	662·93	...	...	127·06	
17	1st	7·54	N.	E.	436	431·85	...	59·92	...	
18	1st	12·15	N.	E.	94	91·86	...	19·95	...	
19	1st	7·44	N.	E.	216	214·04	...	29·07	...	

1718·62 145·79 388·62 1637·50

Subtract^g. lesser from greater 145·79   388·62

Nett result 1572·83 and 1248·88

COMPUTATION. No. 3.

20	1st	62·9½	N.	E.	655·3	306·04	...	579·43	...	

Add Northing 1878·87   Subt. Easting } 669·45

the angle *on whose line in the Tables* should have been sought the length of the third line; and not on that of its complement 19° 37': this should be carefully remembered in these opera-tions. It is perfectly true that the "line" in the Tables of an angle and its complement is always *the same*, but the *designation* of the two co-ordinates are in one case the *reverse* of those of the other.

The preceding traversing operation No. 1, embraces the lines of the plotted example of a survey at page 30 from *A* to *B*, lines 1 to 10; the No. 2 embraces the left-hand series of those lines from *A* to *C*, and diverges from the right-hand series at (2) 78.

On comparing the No. 2 with No. 1 computation, it will be seen that the No. 1 line is identical in both, being in fact the same line; No. 2 line of No. 2 computation is so much of the No. 2 line in the No. 1 computation as lies between the commencement of the line and the point of branching off, viz. at 78. From this point they are quite a distinct series of lines. No. 15 line goes due west, and it has therefore no pair of co-ordinates, but the *whole length* of the line is placed *in the column " Westing."*

The nett results will be found to differ from the results of No. 1 operation in having a less Northing, and a greater Westing; and a little reflection will shew that the co-ordinates of the "point or angle" to be given at *C*, to guide a drivage to *B*, will be a Northing and an Easting; consequently it will be in the first quadrant.

No. 3 computation, line 20, supplies the data to make the results of No. 2 equal to those of No. 1, arriving in fact at the same point; and a comparison will shew that the Northing in No. 1 is 306·04 more than in No. 2, and that the Westing in No. 1 is less than in No. 2 by 579·43; and we must reduce the No. 2 Westing to that extent, by having an Easting co-ordinate; therefore a Northing and an Easting: from these data we must now decide the angle and the distance, according to the method before given: viz.

Rad. + Log. 306·04 = 12·4857782

Log. 579·43 = 2·7630010

9·7227772 = Tan. ∠ 27° 50½.

Therefore the angle opposite the co-ordinate Base is

$90^\circ - 27^\circ 50\frac{1}{2}' = 62^\circ 9\frac{1}{2}'$; and taking this ∠ and the co-ordinates as follows.

		Per.	Base.	Numbers.
		306·04	579·43	
and finding		280·22	530·54	= 600
	rem.	25·82	48·89	
,, ,,		23·35	44·21	= 50
	rem.	2·47	4·68	
,, ,,		2·33	4·42	= 5
	rem.	·14	·26	
,, ,,		·14	·26	= ·3

we thus find the length of line No. 20 equal to 655·3 links.

We might have arrived at the same result if we had taken the whole of the lines around from *B* to *C,* and have combined them in one computation, omitting No. 1 line and the first 78 links of No. 2; that is, taking No. 2 line as 276 links instead of 354: in that case however either the right or left-hand series of lines would have *a different designation,* accordingly as the survey was taken from *B* to *C,* or from *C* to *B*; if the former, the *right-hand* series would have opposite designations, and if the latter, the *left-hand* series would have the opposite designations. This operation may easily be performed as an example for practice.

Having gone thus fully into the working of underground traverses, and there being really no difference in the mode of application of these principles to any lines either on the surface, or otherwise, there will now be no difficulty in applying them to any series or combination of lines whatsoever to which they are applicable. It is unnecessary therefore to pursue this part of the subject any further.

CHAPTER VIII.

AN IMPROVED DIAL AND PROTRACTOR.

For the adoption of the true meridian as the working base-line of the Survey and Plan, it is necessary to have an instrument suited to such method of working; for the practice of the process of computing it from the "declination", through all the varied angles of each successive survey, and thereby reducing the several angles to their relation to the true meridian, would be a matter of great labour, and its practical difficulty would, as a rule, prevent its adoption; and the practice of a more correct process thus requires an improved instrument for the direct taking of the angles of a survey, in relation to the true meridian as a base-line or zero.

It would be easy enough, in the matter of a simple surface survey only, to represent upon the plan the true meridian; but in the matter of mining surveys where the Plans are being constantly added to over a long series of years, and where the magnetic needle is the medium (and doubtless the best) of adjustment for the instrument, it seems to be essential, to the adoption of the true meridian as the working base-line, that the instrument itself should contain the appliances and means for the necessary correction; while, at the same time, admitting of the same simple method of reading the angles as in the ordinary make of instruments.

To provide for this we have designed an improved circumferentor, by the use of which the correction is made without in any way complicating the method or operation of working the instrument, or of reading and entering the angles of the survey. This improvement consists of an adjustable vernier, working in the divided ring of the dial-box to which the needle is set, and on which the reading verniers move; and it reads on the inside of the ring, and downwards to the bottom plate of the dial-box: and a small portion of the initial part of the fourth quadrant is divided on the said plate to half-degrees for this operation; and on this the vernier, which reads to minutes, is adjusted; so that its reading, reckoning from the ordinary zero or 360°, expresses the correct declination of the magnetic meridian for the time being. This adjusting vernier carries a small angle piece which works flush with the inside and top of the ring, and for which a small portion of the ring is sunk, by cutting away; and the divisions so cut away from the top of the ring are recut on the sunk part, so that the instrument may be used by the old methods when desirable or necessary: the small angle piece carries a line, in prolongation of the zero line of the adjusting vernier, on its side and top, and to this line the needle is always set instead of to the ordinary zero of the instrument.

The needle being set to this adjusting zero, the ordinary zero of the instrument will be in the *true meridian*, and all the readings of the survey will express the several angles which the successive lines make with that meridian.

If then the meridional lines laid upon the plan be those of the *true meridian*, the protractor must be laid upon those lines; and the "readings" are to be then taken, or marked off, just in the same manner as though the magnetic meridian was being worked from—there being in this respect no difference in the process; and all computations are thus rendered unnecessary.

It will be necessary to adjust this "declination" or adjusting vernier, either quarterly, half-yearly, or yearly, according to the necessities of each surveyor; but in general it would be sufficient to make the adjustment half-yearly. And the plan should have the dates of the general surveys, and a record of the amount of the declination to which the instrument is successively set, entered in tabular order regularly, so that at any future time the later surveys may be made precisely to accord and compare with those made at former periods.

The adjustment is easily effected by the use of a small key, by which the vernier can be moved either way at pleasure; but it is not liable to get out of order, and will not move except when acted on by the accompanying key.

For the purpose of applying this principle to surveying it is necessary to have the means at hand to correct the instrument periodically to the true meridian. This may entail some little labour and expense at the first, but surely the desirability of having first-class Plans will far outweigh any such consideration. Only those who have been much engaged in mining operations, and have observed the great expense very often attendant on bad surveying and bad Plans, as well as the frequent loss of no inconsiderable quantity of valuable mineral, will for a moment question the desirability of adopting good and correct methods of surveying,—unless indeed such persons are blinded by prejudice against the introduction of improvements, in this, as well as in other matters connected with mining.

The expense of procuring the improved instrument, or of having the improvement applied to an existing instrument, is indeed inconsiderable. For the application of the principle continuously upon any property it is desirable—though not *absolutely* necessary—to have two distinct points of observation, either *corresponding with* the true meridian, or making a *known* angle with it. This having been procured it only remains to take the instrument periodically to one of those points, and observe with the adjusted telescope the pole or other object at the other point : if they are *in* the true meridian, set the zero and telescope of the instrument to the said pole, or, if they make a *known* angle with the true meridian, set the verniers of the instrument to the said required angle, and turn the entire instrument round towards the pole or object until the telescope accords with it, while the verniers read the permanent variation of those points from the true meridian. Then observe if the needle and the adjusting vernier agree with each other ; if they do not, take the key of the adjusting vernier and move it until it accords with the needle exactly, and the readjustment is complete. Care must be observed in performing this last operation to guard against any attraction of the needle from any cause, or with the correction of one error another will be committed ; proper diligence will however ensure against this.

Now it is not absolutely necessary that the true meridian should be laid out upon the property, in order to practise this improved method of surveying, for we may take *any* line near the true meridian (so that it is distinctly understood), and adjust from that line periodically, in the same way as before mentioned, and so ensure the advantages derivable from a correction for magnetic variation: it is doubtless however preferable to have the line laid out either *in* or *from* the true meridian, so as to apply the principle more completely.

We give subjoined two methods for finding the true meridian at any place whose latitude and longitude are known, or are determined.

1st. By equal altitudes of the sun, or a fixed star.

The theodolite being fixed in a position selected as one of the points in the desired line, measure the horizontal angle from the said line to the sun or fixed star (whichever is used), when it is not near the highest or lowest point of its apparent daily course, and take also the altitude, adjusting the intersection of the hairs of the telescope to it, in some marked and definite manner. Leave the limb clamped and let the instrument remain quite undisturbed until the sun or star is approaching the same altitude at the other side of its apparent circular course. Then, without moving the body of the instrument, unclamp the limb, and direct again the telescope towards the same object, and follow it with the aid of the tangent screw until the cross hairs correspond precisely as before with the same relative position upon the disc of the object, as in the first observation: this being done, read off the horizontal measure of the angle between the assumed line and the new direction of the observed object; the mean between the two horizontal angles will define the position of the true meridian, *if a star has been observed.*

If the *sun* has been observed a correction is required in consequence of the sun's change of declination. When that declination changes towards the north the approximate direction of the meridian as found by the foregoing method is too far to the right; if declination changes towards the south it is too far to the left.

The required correction is found by the following formula:—

$$\frac{\text{change of sun's declination}}{2} \times \text{sec. latitude} \times \text{cosec.} \frac{1}{2} \text{ angular}$$

motion of the sun between the observations.

Example. Given:—Date of observation (say 26 July); sun's declination changing towards the south at the rate of 13' per day, or 32"·5 per hour; time between the observations, say 6 hours, change of declination for 6 hours 3' 15"; angular motion of the sun between the observations, say 80°; latitude of place of observation, say 51° 34' N.

Then the correction becomes

$$\text{Log. } \frac{3' 15''}{2} = 97''\cdot5 = 1\cdot9890046$$

$$\text{,, } \text{ sec. } \angle 51°34' = 10\cdot2064865$$

$$\text{,, } \text{ cosec } \angle 40° = 10\cdot1919325$$

$$\text{,, } 244'' = 2\cdot3874236$$

therefore the approximate direction as found by the observation is too far to the *left* by 244", or 4' 4",

In the preceding process it must be observed that the mean of the two horizontal angles is their half-sum when they are at the same side of the station line, but their half-difference when they are at opposite sides of the said line.

2ndly. Where Greenwich mean time is given or obtainable, by the aid of the telegraph, procure the exact Greenwich time on some convenient day, and with the "equation of time" change that mean time into true apparent solar time : then take the longitude of the place of observation, and add for each degree of difference in west longitude 4 minutes of time, and for each minute of longitude 4 seconds of time; if the observation is in longitude east of Greenwich the same has to be subtracted instead of added.

Then these differences for longitude having been added to, or subtracted from, the Greenwich apparent solar time, gives the apparent solar time at the place of observation, and at 12 of that time the sun is due south.

A chromatic object-glass for the telescope is necessary in making these observations.

The small round legs of the mining dial or circumferentor are not sufficiently strong for out-door work in windy weather, when the telescope is being used; for the action of a slight wind will often be sufficient to cause the needle to be unsteady. There seems to be no practicable remedy for this but to have a strong tripod-stand for use in out-door surveying; and while this entails the expense of two sets of legs for the instrument, the advantage gained is often much more than sufficient to compensate for the extra cost. The tripod-stand cannot be used with advantage underground, in many places, for it is too heavy and unsuitable; while it is by far the best stand for surface operations.

Improved Protractor.

We have also designed an Improved Protractor, in the using of which the successive lines of the survey are run off at once from the instrument to their positions upon the Plan, instead of making punctures in the paper, with pencil marks and numbers to distinguish one angle from the other; the pricking of the paper is also thus avoided.

The instrument is made by lengthening out the central bar that carries the verniers, to about 3 inches each way beyond the outer side of the divided circle, and at each end of this bar there is a transverse bar: and on the ends of these transverse bars and reaching from one to the other, are fixed two ebony or metal bars—one on each side—sufficiently distant apart for the inner edge to just work clear of the divided circle. The ends of the central bar each carry a small glass with a + cut, so that those two and the central + are in exact alignment, and form the zero of the instrument. Each side of the frame is exactly parallel with this zero line.

The outer side of the divided circle has a projecting shoulder cut on it, and is provided with a clamp, and adjusting tangent screw, as in the Theodolite.

In plotting a survey from this Protractor it is first laid

upon the meridional line in the most convenient position, and each line of the survey is then laid down in succession as taken from the instrument, either with the large set-square, or parallel ruler, and at once laid down in its proper position; and thus the work of first marking off and numbering the readings is rendered unnecessary.

With the large set-squares either one side or the other of this frame will be sure to be convenient for plotting from.

These instruments can be procured from Messrs Troughton and Simms the eminent Opticians, who have made them from the designs of the author.

CHAPTER IX.

PLOTTING THE SURVEY, ETC.

IF the directions respecting the proper seasoning of the paper, laying down the meridional and transverse lines, &c., have been complied with, and the survey made and duly entered in the field-book, the plotting operation will now be proceeded with.

Unless the position of the boundary, and other objects on the surface, can be so far imagined that the site of the pit, or level, can be with confidence fixed for some particular point on the paper (this can frequently be done from some old plan, where such is in existence), the surface survey should be carefully laid down first; and the position of the pit or level having been marked, the underground workings can then be taken in succession.

Taking the underground survey, and the position of the starting point being known or fixed upon, the first thing will be to fix upon an intersection of base-lines from which conveniently to take, or mark off the compass-readings correctly on the paper with the aid of the Protractor.

To produce satisfactory results the surveyor should be possessed of instruments of the best kind, and should then take care of them. Good instruments once possessed and taken care of will last a lifetime, and will always have the confidence of the owner; but even the best tools, if roughly and carelessly used, cannot be expected to ensure correct results; and badly made instruments can never be reasonably supposed to do so : and the surveyor may rest assured that should his plans prove to be incorrect, the faults will never be attributed to his tools, but to his own want of skill, for he will be expected, or supposed, to have correct tools to work with.

A good protractor is a very essential thing, and those with double folding arms with verniers divided to minutes (when of the best make) are, next to the improved instrument spoken of in the preceding chapter, the best; this instrument having been correctly fixed on the meridian line, the center coinciding with one of the intersections, the two readings of each line should be made to coincide with the verniers, with the aid of a magnifying glass, and both points marked on the paper with a small pencil circle around the punctures, and with the number of the line to which they refer, or run off at once to their position, as the case may be. This having been done with the whole of the lines (if with a folding arm protractor), or of as many of them as can conveniently be marked off from that part of the paper, the readings should then be taken backwards in succession as a check, to see if each pair has been correctly marked off: this being done the protractor will now be removed, and the laying down the lines proceeded with.

For running off the lines either from the improved instrument, or from their markings as before mentioned to the position they are to occupy on the Plan, the best instruments are a pair of large pear-tree angles or set-squares : these will be found to be far preferable to a parallel ruler, or even to one set-square and a straight-edge; for the large set-square has a much broader bearing on the paper than a straight-edge, and is much steadier and less liable to those small lateral movements which will frequently be found to occur with the straight-edge (unless it is a large steel edge, and then it is too

large to be handy for this operation), when sliding in close
contact the one over the other, to the position for the lines.

It will be found conducive to accuracy to mark off no more
compass readings in any one place on the paper than can
easily be run off to their position with one (except in occasional instances) movement of the set-squares ; for it will be
better to fix the protractor in a new position, and mark off the
readings nearly on the part of the paper the lines are to
occupy, than to mark off all in one place, and be under the
necessity of running them off across a considerable breadth of
paper, to their several positions; and with the meridional and
transverse lines before recommended, the protractor can be
put at any intersection of those lines desirable.

The first line should now be marked in its position by
placing the side of the set-square to the two angle-points
marked 1 (and a check upon this operation will be afforded
by running the eye along the side of the set-square, and observing if it cuts the intersection of the meridional and transverse lines), then placing the other set-square to that side of
the former which will the most readily permit the proper side
of it *to come upon the starting point* run the line off to its
position : a faint pencil mark should now be drawn along the
edge of the set-square through it with a hard pencil, so cut
that it shall have a thin wedge-like edge, and a flat side to
slide against the set-square ; and having drawn a line sufficiently long to embrace the distance to be marked off, run the
set-square back again to the points to see if it still remains
parallel with them ; this being proved, next take the compass
or dividers with the adjusting leg and the diagonal divided
scale, and set the legs to the length required; or if the line
be too long to be marked off at once in this manner a portion of the length may be marked off with a scale, and the
remainder, including the part less than 100, taken in the
compass and added to the former part; a small puncture
should be made to indicate the end of the line, and it should
be distinguished by a small pencil circle and numbered ; also
remember to put the number at the *beginning* of the line, as
shewn-in, and to correspond with, the field-book ; the next and

succeeding lines should then be operated on in the same way, until they are all marked off in their respective positions. In progressing forward it will frequently be found that the next line to be put on runs almost at right angles to that last plotted, and except the workings of the mine are thoroughly well known to the surveyor he will be at a loss to know whether it should be a turn to the right or left hand, if he has not remembered to indicate the turn in the field-book; and for this reason the practice should never be omitted, but by turning to the field-book the mark should at once indicate which way to draw the line, whether to the right or left hand; and so on of all the other turns that are capable of misleading without a distinct mark for guidance.

Having laid on all the lines, and being assured of the correct lengths, etc., the different objects and particulars entered to each line in succession should now be placed upon the plan, as indicated by the entries in the field-book.

When we come to a point where the survey has subsequently been continued in another direction, as for instance at 78 ③ of the example of a mining survey before given, the length 78 should be marked with the compasses from the diagonal scale, and the "ribs" sketched with the pencil on the proper side, to indicate the position of the commencement of the subsequent work; but where stall roads, faults, commencement of gobs, ribs, or other things have to be indicated it will be sufficiently accurate to mark them off with a pencil, from a long scale laid upon the line in a correct position for so doing, or with a small set scale where necessary. The matter to which these lengths refer should be marked down to each line in succession (after all the lines themselves have been laid down), as the work of filling up proceeds, so that there may be no error from memory in trying to remember subsequently to what the marks refer.

In this way should all the workings of a colliery, or other mine—with the exception of stall-points—be marked down in succession; and as each page of the field-book is completed a mark should be made upon it distinctly, to show that it is done with.

In reference to the points of stalls, it will be unnecessary to lay down these points in the same manner as those of the important headings, and other parts of the workings. The handiest way of laying down the points of stall roads is to take the small protractor which usually accompanies the other drawing instruments, and laying it properly on the meridional line most conveniently situate, and at one of the intersections, mark off the point with the pencil, and at once with the parallel ruler run off the line to its proper position, and lay down the stall to which it refers ; and so on of the others in succession. This will be sufficiently accurate for these things, and they will be pricked off and laid down at once in the operation of laying down the details, instead of mixing them up with the points of the headings and other more important parts of the survey.

If the mine "pitches" too much to make the correction to reduce the measure to "horizontal" as the work of surveying proceeds, care should be observed to make the reduction before the plotting is proceeded with. Where the pitch is general, one taking of the angle of elevation or depression will mostly be sufficient ; but if the pitch varies, it will be necessary to observe the angle of elevation or depression more often, and to enter each distinctly : if it is considerable it will be advisable to reduce the lengths of the whole lines, and of any intermediate particular object by the Traverse Tables, for those tables are as applicable to the reduction of hypo. to base, as to the working of traverses on the flat. But here it must be borne in mind, that the horizontal measure will *invariably* be represented by the figures as found in the column "latitude" or "perpendicular," and the other column will show the vertical rise or fall in the length of the line : in this application of the tables they form a ready medium for correcting the different rises in mines that pitch much, and by an expeditious and correct operation to facilitate the elimination of the true horizontal measure. And where the pitch is considerable, 10° or upwards, the horizontal measure decreases rapidly, and renders it requisite to observe the more care in obtaining the angle of altitude.

While the correct laying down of mining surveys is not difficult to those who go about it with care and diligence, and with a genuine intention to ensure accuracy, it will seldom occur that carelessness or inattention, or a despising of those small niceties which are essential to ensure precision in working, will result in anything but annoying disappointment.

The whole of the workings having been put upon the plan, together with all the details of which note has been made in the field-book, the matter of "inking in" will be ready to be proceeded with. Where two or more veins of mineral are worked one over the other, and are placed upon the same plan, it will be necessary to distinguish one from the other in the colours assigned to the several separate workings, according as fancy or taste may decide, so that each may easily be recognized, with their several connections. Some use the brush in distinguishing the veins one from the other, using black to mark all the workings in, but painting the spaces of each between the lines with a distinguishing colour: of course, these details are very much matter of taste or opinion, and are not of such importance as those principles upon which *accuracy* depends; but we hold that, so far as possible, it is better to dispense with the brush, and use the colours only with the drawing pen; and we consider that a mining plan will generally look neater and better managed in this way than if the colours are put on with the brush, over all the exhausted space. With a surface plan which is finished at once the colouring, shading, etc. may be put on with the brush, and will if done well look well; but with mining plans which are constantly being *added to* it is different, for in this case if the shades as laid on with the brush are not precisely alike they do not look well; and if the colours selected are not simple colours the difficulty of exactly matching them is increased, therefore we think that as much as possible the brush should be dispensed with on a mining plan.

It is very easy in competent colouring hands to make inaccurate plans *look* well, and it is easy to deceive the inexperienced into saying much in praise of a plan so finished, and of the person who has finished it; but no amount of colour

E

will make an *accurate* plan : and while every surveyor should seek to finish his work neatly his chief care should be to be accurate, and this can only be attained by constant attention to, and practice of, those essential principles upon which accuracy depends.

The manner of laying down the surface lines, and the details of the surface survey, does not essentially differ from that already indicated for the underground survey. In general the surface lines will be much longer and less numerous than the underground lines, and their accuracy will admit of frequent proof. If the groundwork of a survey has been formed by a series of triangles and long base lines in the way indicated in Chapter XI., the laying down the details and the filling in the shorter lines will be comparatively easy ; and by a proper disposition of those lines to easily embrace the fences etc., and distinct memoranda in the field-book shewing where they abut on the main base lines, their accuracy, and the accuracy of the plotting of the main lines, will be constantly and reciprocally checked.

The finished plan will of course bear the descriptive title of the property ; the name by which it is distinguished ; together with the parish and county in which it is situate, and the name of the owner ; also on the different boundaries the names of the owners of the adjoining properties. The scale to which it is made will be indicated, and also the detailed area of the property where necessary, and the total area. It will also bear the names of any particular farms, houses, brooks, rivers, woods, faults, or other objects contained within the property.

If the paper has been divided into 10 acre squares as before indicated these will generally assist in the computation of the area : if the whole squares are taken separately, and the parts separately, and the latter added to the former, the aggregate will be the total area. It will however be well to take also another method of computation, that of computing the property as a whole, without reference to the squares ; and by comparing one result with the other a check will be obtained upon each computation, and the result obtained will be entitled to greater confidence. The area will of course be temporarily divided for computation into trapeziums, triangles, etc. principally, and the rules of practical geometry applied to compute the area of those and other figures necessarily entering into the division.

If the meridional lines upon the plan are those of the True Meridian there should be a columnar table on one margin of the plan, in which should be successively entered the date of the last general survey, and also the amount·of "declination" to which the instrument was adjusted for the survey; this will then afford a systematic view of the correction continuously applied, for reducing the angular readings to the True Meridian.

It will be found very useful to have a sheet of good tracing paper or tracing cloth, divided into one acre squares (drawn to the same scale as the plan with which it is to be used), in black lines, and each of these squares bisected in both directions with ink of another colour in faint lines, which subdivisions will consequently be each equal ¼ acre. If it is then necessary to see approximately the area of ground exhausted of any particular vein of mineral, it will be only necessary to lay this sheet of divided paper on the exhausted space represented on the plan; and by counting the whole acres, and counting and averaging the ¼ acres (which can be done by a practised eye very nearly), and adding these together, the area will be obtained to very near the truth, without using pencil lines about or drawing figures upon the plan. And by doing this at the end of each year, and comparing the area of ground exhausted with the coal or other mineral obtained from it, it will be seen at once whether the vein is producing the quantity of mineral which under the concurrent circumstances it ought to do. The side of a square whose area is equal one acre measures 316·228 links.

As examples of the quantities we have proved different veins of coal to produce per acre, for each 1 foot of thickness of the veins, we beg to refer the reader to what is said on that subject in Chapter XII.

CHAPTER X.

THE BUSINESS AND APPLICATION OF SURVEYING.

THE work and business of a surveyor enters largely into mining operations, and particularly into coal mining, and it may very fairly be said that the extent to which good surveying is applied in mining, goes in some degree to indicate the advance made towards a perfect system of management; for although it can never take the place of, or supply the default of, discipline and good scientific knowledge, yet the application of scientific knowledge will proceed, in no small degree, through the channel of good surveying.

Given: A mineral property to be worked, and the direction in which the measures dip being known, it will be of the first importance to know the figure of the property under consideration, so that the position for the "winning" may be determined upon, in order that it may be so situate as to facilitate the economical and convenient raising and draining of the minerals, as well as their despatch to market; and thus a correct Map will be essentially necessary, so that the plan of working may be determined upon. It is still however often the practice to allow the colliery or other mine to form itself, as it were; there being no prearranged plan of development, but simply leaving the different principal headings to follow what appears to the workmen the most natural course. This plan however generally, or rather *invariably* leads, sooner or later, to a complication of workings ill adapted to ensure either the most perfect ventilation of the mines, or the convenient and economical carriage of the minerals to the bank.

If it is a winning by means of a pit, and it is in contemplation to raise a large daily quantity, the laying out of the pit bottom will be a matter that should be well considered; for as this part of the mine will have to pass the whole daily quantity of mineral raised, economy is sometimes considerably affected by the way in which it is laid out: and the most perfect arrangement for passing the full wagons on to the carriages, as well as the empties off for entering the workings, will be found to be worth all the thought that can be bestowed upon it, so as to ensure convenience and despatch. Where the empties are conducted away by a separate road, which has just sufficient fall to keep them moving on to where they are taken off by the horses, or other hauling power, the road will require a fall of from an half-inch to an inch per yard, according to the state the trams are kept in for easy running.

Engine drawing planes on the levels, self-acting incline planes to the rise, and engine drifts to the dip, are now being more generally used than formerly, to cheapen the cost of haulage by reducing the number of horses, as well as at the same time to keep the air more pure and cool than it can be where so large a number of horses, as some collieries would otherwise require, are passing through all the workings continually; and the correct and well-considered laying out of these works and their "banks" will have a considerable effect upon the subsequent cost of working them; for if they are not (except in regular and intentional curves) practically straight, the wear and tear and consequent cost of the wire ropes are very much increased; so that it becomes a matter of great practical importance to get the engine and incline planes in perfect alignment, thereby to ensure that whenever the ropes are by the strain lifted off the rollers they shall again fall back into their correct position, to prevent chafing and cutting against the ground, and surfaces that are not rolling: and if the supervision of the opening of this work is given into the hands of the surveyor, he should take particular care that it is thoroughly done.

There is again the work of opening the headings so that they may be as near as is practicable parallel with each other, when the best point for them has been determined upon; for few things look worse on a mining plan, or prove more objectionable in the practical working of mines, than having all the headings going at random, some to the left hand and some to the right, some giving the stall roads 40 yards to go, and others giving 140 yards before they reach their end; but if they are driven parallel the coal can be worked much better, and more systematically, and as a natural consequence cheaper and cleaner. And while we know that it rests with the management as to whether such a system is allowed to obtain, and that it cannot well exist where there is a systematic plan of having "points" put on all the headings (and the workmen made to keep them), yet there is very frequently much less care bestowed on this matter than there should be, even with those who will at once admit the necessity and desirability of the principle.

It is sometimes the practice to have points put upon all the stalls where "long-work" is not adopted, as well as upon the headings, and in some veins there is no expenditure of labour that returns a more ample interest than does this; for it is of the first importance to keep the pillars their proper thickness while the roads are going in, to prevent their being reduced, and often as a consequence destroyed, by being "robbed" and made too weak to sustain the top—occasioned by crooked driving; and not only do the pillars thus get crushed and rendered partly, and occasionally wholly worthless, but heavy falls are often thereby occasioned, and sudden outbursts of explosive gas rendered more frequent, which would have exuded more gradually, but for the fall, and the ordinary currents of air would have been able the more effectually to dilute it, and render it less dangerous than is possible when it comes off in large quantities suddenly. Where long-work is adopted, there is not so much necessity for points on the stalls, as there is a much better chance of keeping the roads nearly parallel, when they are open one to the other along the face.

In the making of quarterly or other periodical surveys, it is of importance to have a distinct knowledge as to the position at which to begin. Where it is possible to mark each survey, by placing a *D* or any other mark determined on upon the rib or roof, there is not much difficulty in the matter; but it is not often that the positions can thus be easily determined, and then the surveyor has to fall back upon his knowledge of the relative position of some heading or stall parting, or other object, and trace out his last terminal position from that. Where there is not much changing of the workmen the name of the man who was working the last road in a panel of work taken, might be entered in the proper position in the book, and the chainage of this parting "cross" being taken it will generally be a good mark from which to define the position at which a former survey terminated, and from which to continue the work of surveying at a subsequent period : though this is, generally speaking, a very correct way of proceeding, yet at times there is a possibility of the surveyor being misled by having the wrong road pointed out as that for which enquiry is being made ; and hence constant vigilance will be essentially necessary to ensure against error in this respect. Sometimes it is necessary to measure a level or other heading from some distinct turn, or other conspicuous object ; and the system adopted, and regularity with which a surveyor is able to select these points and mark them in his book with clearness for subsequent reference, will greatly influence the time that he will be occupied in going around a large mine to take the new work opened ; while an uncertainty in having positions to start from that are but ill defined in the field-book, will often lead to error by the operator taking a wrong point of departure, instead of spending the requisite time to ascertain the correct: hence nothing will contribute more to accuracy than a thoroughly clear and careful method of keeping the field-book, and of entering in a uniform manner the different matter which pertains to the survey made, and whatever may forward the work of continuing the survey at a *subsequent period.*

To facilitate the work of periodical surveys, we would strongly recommend the surveyor to adopt some particular *route* through the workings, and to keep up that route at all his surveys. Also, before going about the survey of a mine or colliery at which he has been before, to sit down and go over all the principal points of his last survey,—making notes in succession upon the different points from which he is about to continue the work of former surveys ; these notes being entered in succession in a condensed form on the first page allotted for the new survey, will be all under his eye, and he will

see them all at a glance when proceeding to the operation of
surveying, instead of having to search through the whole of
the entries made on the last occasion, at each different point
of the work. Much time is thus saved in the underground
operations, and this system has the additional advantage of
bringing before the mind's eye the whole of the workings, and
may perhaps recall to mind several matters of importance
which it is advisable to note closely, and which, but for it, would
probably have been overlooked. It also induces a correct
habit in the matter of making entries in the field-book, and of
so placing the particulars that they shall be uniform for extracting the notes when required on the next occasion.

The matter of moving points and setting out the gradients
of drivages has been treated of in another part of this work ;
therefore it is unnecessary to go further into that part of the
subject in this place. The learner will have no difficulty in
ascertaining the point to be given to a drivage, if the line to
be followed be first laid down in position on the plan, and
next with the parallel ruler or other instrument it be run off to
an intersection of the meridional and transverse lines; after
which, placing the protractor in its proper position, see what
is the angle made by the line of the proposed drivage, and
making a note of the same in the field book, give that angle
with the instrument, or dial, to the proper place underground.

In laying out the siding communication, and railways, to
the tips or screens of an extensive colliery, or for Iron Works
or other operations of large extent, there is need of great care
and discernment, and few things will more effectually test the
skill of the surveyor, than the designing and planning of these
things—indeed it may be almost considered as an entirely
distinct matter from the ability to make an ordinary survey
and plan; for the latter is, more simply, the making a plan of
things, as they already exist, while the former is the *designing*
the means of communication for an extent of operations which
demand convenience of arrangement with facilities for despatch,
in all its details, and this will call for the exercise of sound
engineering skill; and upon the successful manner in which
these are laid out to meet the peculiarities of the traffic to be

accommodated, and provided for, will depend the amount of engine or other power that will be required to move about, and to place the same; and it is not too much to say that a bad arrangement of roads will speedily involve the employment of an extra engine for shunting, with all its attendant expense, besides perpetuating a continuous existence of more wear and tear, upon both roads and trucks: therefore, the accomplishment of this part of the surveyor's or engineer's work will demand his best energies, for the successful accommodation of the traffic of large works.

It is of great importance to arrange the gradients, wherever it is practicable, so that the wagons shall pass under and away from the colliery screens by their own gravity, giving just sufficient fall to the road to overcome the friction of the rolling surfaces, and thus avoid the necessity of having horses always in attendance: and while in many places the natural conformation of the ground is such that this can be accomplished without the least difficulty, there are others which require to be carefully thought out, to properly accommodate the requirements of the works. And as the rails of the sidings about works are seldom very clean, consequent upon the nature of the traffic to which they are subject, it will not be advisable to count upon flatter gradients than about 1 in 100 for the wagons to run alone, and if a little more fall can be given so much the better; for while a gradient of 1 in 100 will allow for about 22·4 lb. per ton for friction, and should under good conditions be much more than sufficient, yet if wagons and roads are not kept in good running order there will not be any considerable superfluous margin.

In Iron Works it is often impossible to arrange the roads with falling gradients in either direction, to in any appreciable degree facilitate some portions of the traffic; and occasionally the nature of the traffic less requires it than at collieries; but still in such instances there is always much to be considered in the disposition of the roads and partings, to provide for that convenient despatch which is essentially necessary, where so much is being done constantly and unremittingly, night and day; and for which no further rule can be given than that each particular case must be considered on its own merits, always bearing in mind the quantity and kind of traffic to be provided for, together with the space at hand that can be assigned for its accommodation: and wherever an advantage is to be gained by the adoption of some particular gradient, and there are no insuperable difficulties in the way of its being carried out, such details should always have due consideration.

The practice of levelling for mining purposes does not much differ from that of taking levels for other works, further than this, that when it has to be done underground it is accomplished under very considerable disadvantages; and great care is necessary in throwing light into the telescope, so as to be able to see the cross-hairs, and to get a steady light, yet only what will be just sufficient for the purpose; and also to hold the light as much as possible in the alignment of the axis of the telescope, or else the cross-hairs will appear very unsteady, and the vision of them will frequently be flickering and uncertain to the eye. A good steady light will also be necessary for the assistant who holds the staff, so that the figures may be seen with ease and certainty. There is often great difficulty in securing anything like a good B. M. (bench mark) underground, in consequence of the movements of the ground, but the inconvenience arising from this is generally much less than in important surface works.

The business of surveying as applied to other operations than that of mining etc., has been so frequently and ably treated of before by others, that those who seek specific information on such branches may safely be referred to those works for the information they are in need of.

In the succeeding chapter will be found a more full explanation and illustration of the application of Triangulation to a considerable surface survey; and as the value of a Map depends upon its accuracy, and in mountainous districts the operation of chaining is rendered much more difficult and less accurate, by reason of the uneven nature of the ground, it seems to be most desirable that the principal groundwork of every important survey should be formed by the application of those principles, therein explained and applied.

CHAPTER XI.

ON TRIANGULATION, ETC.

WE will now proceed to give an example of the application of Trigonometry to a surface survey, shewing how a property of a certain figure may be divided for Triangulation, and computed through the whole series of lines. This example is not from any actual survey, but is given as an instance of the application of the principles enunciated in the foregoing chapters of this work.

The superiority of these principles—if accuracy is considered to be at all desirable—cannot be for a moment doubted, and the slightly increased amount of labour, if indeed there is any increase of labour, is abundantly compensated by the increased value of a thoroughly reliable Plan on which to lay out the workings, and guide them for the proper exhaustion of the minerals throughout the property to the boundaries, and the reservation of proper barriers to ensure against flooding and other mischiefs.

In the example here given, Plate 3, line *a* of No. 1 triangle is that supposed to be measured very accurately, for computing the whole of the lines of the Triangulation; and this line, called the "Base of computation," is here represented to be 2564 links. The process of computation is carried on, first from line *a* to line *j*, embracing *b, c, d, e, f, g, h,* and *i*; then from line *a* to line *k*, embracing *q, p, o, e, n, m, l, i,* and *k*; thus the lines *e,* and *i*, come into each of the series of computations, and will each form a check upon both series, one of the other; and if these lines come the same in each (as they ought to do) it is so far satisfactory to the operator. Then, selecting one of the lines—say, for instance, *k*— as a "Base of verifica-

tion," it should be measured on the ground with the chain,
and checked with extreme care, to ascertain if it corresponds
with the computed length: if it comes within ·5, or 1 link of
the computed length, it may be looked upon as being very
satisfactory.

Care should be taken to avoid as much as possible the
adoption of angles greater than 90°, as they are neither so
convenient, nor so easy of computation: it should be also
carefully observed that the sum of the angles of each triangle
equals 180°; and that the sum of all the central angles of
each closed polygon (at *A* and *B* for instance) equals 360°.

An instrument, theodolite or circumferentor, of the best
make, and in perfect adjustment, will be requisite for taking
these angles; and having carefully levelled the instrument the
surveyor should proceed to take the angles, and enter them in
a sketch, made for the purpose, of the several triangles in
their proper relative positions. In some states of the weather
the refraction of light will have a very perceptible influence upon
the operation of taking the angles, and this should be carefully
noted, and errors as far as is possible guarded against: for
instance, if there are passing clouds, and *one* observation (or
one of the readings for determining the angle) be taken while
the sun is obscured and the other *after* the cloud has passed,
there will often be a perceptible difference (dependent upon the
state of the atmosphere) in the measure of that angle, com-
pared with the measurement when both observations have
been taken either in sunshine or in shade. Therefore it is
well and proper to bear this in mind, and to check the readings of each angle by a double operation, before the instrument
is removed from the position or station.

At *A*, and *B*, the whole of the angles at the center of each
polygon can be taken and checked at one fixing of the instrument.

The trigonometrical Rule by which these computations are
made, is that given before:—viz. *Add the logarithm of
the given side to the sine of the angle opposite to the side
required, and from the sum subtract the sine of the angle
opposite to the given side; the remainder will be the logarithm*

of the side required: or, instead of subtracting the sine of the angle opposite to the given side add its co-secant, and the sum will be the logarithm of the side required.

Then, the given side a being 2564 links, we have the following computations.

To find side b.

As sine of $\angle$ 82° 33′ = <u>9·9963183</u>

: log. side $a = 2564$ = 3·4089180

:: sine of $\angle$ 40° 30′ = <u>9·8125444</u>

13·2214624

: log. side $b = 1679·4$ = <u>3·2251441</u>

To find side c.

As sine of $\angle$ 46° 21′ = <u>9·8594804</u>

: log. side $b = 1679·4$ = 3·2251441

:: sine of $\angle$ 75° 34′ = <u>9·9860720</u>

13·2112161

: log. side $c = 2247·7$ = <u>3·3517357</u>

To find side d.

As sine of $\angle$ 75° 34′ = <u>9·9860720</u>

: log. side $c = 2247·7$ = 3·3517357

:: sine of $\angle$ 58° 5′ = <u>9·9288145</u>

13·2805502

: log. side $d = 1970·1$ = <u>3·2944782</u>

To find side e.

As sine of $\angle$ 38° 21′ = <u>9·7927163</u>

: log. side $d = 1970·1$ = 3·2944782

:: sine of $\angle$ 76° 29′ = <u>9·9878012</u>

13·2822794

: log. side $e = 3087·2$ = <u>3·4895631</u>

To find side *f*.

As sine of ∠ 38° 21′	= 9·7927163
: log. side *d* = 1970·1	= 3·2944782
:: sine of ∠ 65° 10′	= 9·9578626
	13·2523408
: log. side *f* = 2881·5	= 3·4596245

To find side *g*.

As sine of ∠ 51° 41′	= 9·8946461
: log. side *f* = 2881·5	= 3·4596245
:: sine of ∠ 68° 35′	= 9·9689262
	13·4285507
: log. side *g* = 3419·	= 3·5339046

To find side *h*.

As sine of ∠ 68° 35′	= 9·9689262
: log. side *g* = 3419·	= 3·5339046
:: sine of ∠ 59° 44′	= 9·9363574
	13·4702620
: log. side *h* = 3172·	= 3·5013358

To find side *i*.

As sine of ∠ 48° 7′	= 9·8718681
: log. side *h* = 3172·	= 3·5013358
:: sine of ∠ 75° 9′	= 9·9852468
	13·4865826
: log. side *i* = 4118·2	= 3·6147145

To find side *j*.

As sine of ∠ 75° 9'	= 9·9852468
: log. side *i* = 4118·2	= 3·6147145
:: sine of ∠ 56° 44'	= 9·9222721
	13·5369866
: log. side *j* = 3562·4	= 3·5517398

To find side *q*.

As sine of ∠ 40° 30'	= 9·8125444
: log. side *b* = 1679·4	= 3·2251441
:: sine of ∠ 56° 57'	= 9·9233450
	13·1484891
: log. side *q* = 2167·4	= 3·3359447

To find side *p*.

As sine of ∠ 46° 6'	= 9·8576648
: log. side *q* = 2167·4	= 3·3359447
:: sine of ∠ 75° 20'	= 9·9856129
	13·3215576
: log. side *p* = 2910·	= 3·4638928

To find side *o*.

As sine of ∠ 75° 20'	= 9·9856129
: log. side *p* = 2910·	= 3·4638928
:: sine of ∠ 58° 34'	= 9·9310750
	13·3949678
: log. side *o* = 2566·6	= 3·4093549

To find side *n*.

As sine of ∠ 68° 8′ = 9·9675728
: log. side *e* = 3087·2 = 3·4895723
:: sine of ∠ 61° 23′ = 9·9434172

13·4329895

: log. side *n* = 2920·2 = 3·4654167

To find side *m*.

As sine of ∠ 46° 7′ = 9·8577863
: log. side *n* = 2920·2 = 3·4654167
:: sine of ∠ 86° 15′ = 9·9990691

13·4644858

: log. side *m* = 4042·9 = 3·6066995

To find side *l*.

As sine of ∠ 86° 15′ = 9·9990691
: log. side *m* = 4042·9 = 3·6066995
:: sine of ∠ 47° 38′ = 9·8685548

13·4752543

: log. side *l* = 2993·5 = 3·4761852

To find side *k*.

As sine of ∠ 44° 46′ = 9·8477091
: log. side *l* = 2993·5 = 3·4761852
·: sine of ∠ 59° 36′ = 9·9357660

13·4119512

: log. side *k* = 3666·4 = 3·5642421

To find side *i* by this series.

As sine of ∠ 44° 46' = 9·8477091

: log. side *k* = 2993·5 = 3·4761852

:: sine of ∠ 75° 38' = 9·9862017

13·4623869

: log. side *i* = 4118 = 3·6146778

being ·2 difference between the first and second computation to find the side *i;* this is satisfactory.

If it should be required to ascertain what the respective outer angles of No. 9 triangle *should* be, so as to close the polygon composed of triangles 3, 4, 5, 9, 8, and 7, and the length of the line *k* between the two abutting triangles 5 and 8, we must then apply the following rule. Having two sides and the angle between them; to find the two other angles and the third side. Rule:—

If the angle included be oblique, add the logarithm of the difference of the given sides to the tangent of half the sum of the unknown angles, and from the sum subtract the logarithm of the sum of the given sides, or add its complement; the remainder or sum will be the tangent of half their difference.

Applying this rule, we have the two given sides = 4118·3 and 2993·5; and 2993·5 + 4118·3 = 7111·8 = the sum; and 2993·5 − 4118·3 = 1124·8 = their difference.

Also sum of unknown angles = 180° − 59° 36' = 120° 24'; and 120° 24' ÷ 2 = 60° 12' = the half sum.

Then, Log. diff. 1124·8 = 3·0510367

Log. Tangent 60° 12' = 10·2420687

13·2931054

Sub. Log. sum = 7111·8 = 3·8519795

Log. Tangent 15° 26' = 9·4411259.

Therefore the angles will be, 1st, 60° 12' + 15° 26' = 75°.

38′; and, 2nd, 60^0 12′ — 15^0 26′ = 44 46′; and the side k will then be computed by the rule given for the other sides.

The diagram of lines of which the foregoing is the computation forms a complete foundation or groundwork on which to proceed to fill up all the details of the estate to which it may refer. Good firm stakes, or some other reliable marks, should clearly distinguish the positions of all the stations, and then poles can be set up for measuring those lines or combining short diverging lines with them, as circumstances may require, in order to take in all the requisite boundaries, fences, brooks, houses, or other matter; and, if found necessary, portions only of some of the lines can be measured from one or other of the ends if complete information for plotting can thus be obtained, and so that the correct particulars are entered in the field-book. In this way awkward and rugged parts of lines, which, with another system, would have to be measured entire to complete the information requisite to plot the work, may be occasionally omitted, without in any manner impairing the accuracy of the Plan : and no error could be continued beyond the triangle in which it arises, and could scarcely arise (with ordinary diligence) without being detected.

The selection of the secondary lines for filling up, and their directions, depend entirely upon the judgment of the surveyor being brought to bear upon the peculiarities and requirements of each particular case, and for which scarcely any two instances will be found requiring the same treatment and division : therefore the better and quicker the judgment exercised in their selection, the more accurate and less laboriously attained will be the result.

The application of these principles is often of great importance also in foreign countries, in the procuring information respecting, and the development of, their mineral resources, and for acquiring a fairly accurate idea of the stratification, correct average line of dip, direction of principal faults, and other information so necessary to guard against error in taking up concessions, and developing the mineral riches of comparatively unknown countries. For while in England, generally speaking, the boundaries of our deposits of coal &c., are pretty well known, in new and undeveloped countries such information has to be gathered.

The object then to be attained, is to procure the data from which to lay down a connected relative view of different and distant points; for while, often, a consideration of the peculiarities and evidences of any one small district, taken alone, will be found to contribute but little to a general knowledge of the surrounding country, and will often indeed be found, when read *by themselves*, to be completely contradictory of the *general* evidences found elsewhere, they will—when read in the light of the *combined* information to be gathered throughout the area under enquiry, and combined also with a consideration of the relative positions of those places, the character, pitch, and strike of the strata as a whole, line of faults, evidences afforded by denudations of the strata, either in large valleys or deep ravines, together with the fossils to be found in the area being examined—enable the intelligent enquirer to form a tolerably accurate idea as to the capabilities of the district under consideration to yield any of those minerals which have been so much a source of wealth and greatness to the countries possessed of them, and of which he may be in search.

In getting up a Plan of such a district as is here supposed, the application of the science of Trigonometry is generally the only practicable mode of doing so; and at the same time it provides a groundwork for the filling in any details in reference to any particular part deemed desirable, either at that or at any subsequent period.

The rules here given are those most generally employed, in such computations as are here indicated; but there are others which have to be adopted occasionally, under peculiar circumstances, instead of these, and which will be found given in treatises on Trigonometry. Also, when the lines under consideration are so large that the curvature of the earth has an appreciable effect upon them, Spherical Trigonometry must be employed, intead of Plane Trigonometry.

While the peculiar evidences of *large* districts must thus

be sought, if coal, stratified deposits of iron ore, and some
other minerals, are being sought after, there are some deposits,
and amongst them the *casual* deposits of iron ore, which do
not become influenced so much by the general features of
large districts, but their evidences are rather to be sought for
in a more confined area. Occasionally an unpromising looking
outcrop will be found to lead to a very valuable deposit of rich
iron ore beneath; while a very considerable area on the slope
of a hill may be reddened by the descent of the rain-waters,
whose solvent action upon a *slightly* disseminated sprinkling of
ore in the strata of the brow of the hill will have brought about
a gradual change of colour, and which may lead some persons
to fancy that a very valuable deposit of rich iron ore is near
at hand. Springs of water holding iron in solution will often
give a colour to the surface over which it flows, and which, if
taken as indicative of what may be found in the immediate
substrata, will be eminently calculated to mislead. We have
taken crusts of brown oxyd of iron, upwards of two inches
thick, from the face of a bank over which has flowed a stream
of water that has contained iron in solution, and which iron
was, most probably, originally taken up from decomposing iron
pyrites in the strata through which the water had flowed, as
that mineral was known to exist near in considerable quantities: and this crust could not have been, from the peculiar
nature of the position, many years in forming. Sometimes the
material of which houses are constructed (if stone has been
employed) will give a clue to deposits in the neighbourhood,
by the occasional appearance of a stray lump of ore amongst
the building material, especially where, as in primitive countries, the people gather and use the loose material they find
lying in their fields, rather than raise it in a quarry. Sometimes the outcrop of a deposit of mineral may be discovered
in going along over a country after a very heavy rain, such as
falls in southern countries, which has washed the surface of all
the loose débris, and left the native strata quite clean and
bare—it being, in fact, nature's performance of that operation
of "scouring," which is still carried out artificially, occasionally,
in many countries, to discover the outcrops of metallic deposits

and other minerals of value. Deep ravines, and the beds of brooks when dry, will also often yield unerring evidence of the presence, or otherwise, of certain metalliferous deposits; as will also the examination of the localities of disruptions, and dislocations of the strata, and also by following the line of contact of rocks which lie unconformably one upon another.

In examining any new country and seeking therein evidence of mineral deposits, these, and many other such-like circumstances, will be made use of by the experienced observer, in endeavouring to ascertain the prospects of success (setting aside for the moment the questions of labour, transit, climate, &c.) of any mining enterprise with which he may be connected and have an interest in establishing upon a profitable basis, and with which the business of surveying must become more or less intimately connected.

CHAPTER XII.

ON THE VALUATION OF MINING PROPERTIES.

As the raising of minerals, and converting them to profitable use, is one of the businesses of life, and not one of its luxuries, except perhaps to an exceptional few, whose luxuries in it consist less of profits derived from the legitimate working of mineral properties than of profits made out of the unwary, by letting them into worthless things at a very high figure under specious colours, and from which no profit perhaps ever has been, nor is ever likely to be made—those who place their capital in mining enterprises do so in the legitimate hope and expectation of reaping a proper return for its employment. There are things in which money is embarked or spent, and in which, if the expectation of profit is not very great, the liability to risk and accident and loss is at least very small; but mining is much too serious a matter to embark in simply from a love of the adventurous, and therefore those who take, or buy mining property, to work it legitimately, and who place their capital in it, have a right to look for an ample return upon their investment.

Whether the question of the manner of deposition of our splendid deposits of minerals, and especially of coal, will ever be decided to the satisfaction of every enquiring mind is extremely doubtful; nor does the value of the possession depend upon the decision of any such abstract question. The fact of its value is unquestioned, and those countries which possess it in greatest abundance, and also the means for its development, are found, taken at large, to be the most advanced in wealth, and prosperity, and high civilization. It is true that we do not always find an advanced state of civilization amongst the workers in our collieries and mines. The miners are a race of men whom some may judge to be somewhat repulsive and uncivilized, simply from the black appearance the coal-dust, or the red skin that the hematite mine and other accompaniments of mining, give them—adhering to the strong and able frame in consequence of the streams of perspiration that pour out through all the pores of the whole body in the laborious work of mining. Such a judgment would however be very inaccurate, and would, as a rule, convey a most uncharitable idea of the miner. That he is impulsive, and often careless of his own life and that of his fellows, we have frequent evidence of in the melancholy testimony borne to the causes which lead, or have led, to the too frequent recurrence of the sad loss of life in collieries and other mines: but it is by no means just to lay all the blame upon the miner when an accident happens, and come at once to the conclusion that it has necessarily originated in his carelessness or foolishness. Other causes, over which he has no control, and for which he is in no way responsible, but by which he is placed in imminent peril, are often operating to accelerate and bring about destruction to life and property; and some of these it is within the range of possibility to prevent, and some are events which it is to be feared no ordinary forethought could have fully averted.

Placed thus in the midst of daily danger, and labouring in the monotonous toil of the mine, it is not to be wondered at if the miner is sometimes found to be heedless and callous; but such is not the case universally, for there will be found amongst miners men of noble impulse and generosity, and some who take a greater notice of the mineral they raise than that merely involved in the raising and filling it into the trams. There are those who are fond of collecting specimens of some of those beautiful fern fronds, and other fossiliferous impressions so frequently to be found in some strata in and about the minerals, which evince an enquiring and teachable mind; and they have also (some of them) formed some opinions of their own respecting the manner of deposition of the veins, &c. they labour in. These, and similar evidences, tend to shew that if the miner is not so highly civilized as he should be, the fault is frequently less his own than of the concurrent surrounding circumstances of his position, and perhaps also of those who are by their opportunities and education placed by Providence in a position that enables them, if they would, to assist in the work of improving his mental condition.

The many works of practical benevolence and love which the miner is found assisting in, and in some instances to initiate, shew that he has a heart as warm and generous as others more generally recognized; and if his heart is warmly appealed to in the matters of religion, purity, prudence, and charity, it will not infrequently be as warmly responded to. Therefore we would submit that if the miner is not in all things what it would be desirable to have him to be, he is not at the same time what he is capable of being made; and those with superior endowments and positions may at all times do much towards raising him to that standard, to which it is very frequently said the "safe" miner should conform.

It is occasionally said that as men grow in intelligence they dislike the mine, and that the possession of one is inconsistent with the performance of labour in the other: but to apply such judgment as a principle, simply because isolated instances may have arisen in confirmation of such an idea, would be manifestly unfair. And can it be for a moment supposed that an All-wise Providence has so placed these great blessings and comforts—as our mineral riches are admitted to be—in a position from which they can be brought only by the prevalence of the most ignorant and degraded class of labour? and that its removal can only be properly secured by the existence of that state of being which is utterly repugnant to, and inconsistent with, those higher attainments of our nature to which, in other works of the same All-wise Providence, we are constantly called? Surely this cannot be, and we should be guilty of great indiscretion if we were to suppose that it is not possible to secure these great blessings, and accessions to our comforts, at a less cost than the perpetual existence of a maxima of vice and irreligion.

Doubtless there will always be enough, and too much, of these in our world while the present course of things continues ; but at the same time there is no reason why a general advancement in intelligence should not be striven after, for the benefit of each generation as it arises ; for the more men grow in intelligence, the less will they become a prey to designing and professional agitators, and the more will the connection between labour and capital be improved. Men will see that it is to the advantage of themselves and of their employers that their labour should be performed with the greatest care, and to secure the best results ; and that if capital cannot be equally well remunerated in mining enterprises with labour, it will not continue to be so employed, but will seek new fields of enterprise for its employment.

Thus we submit that this question of the state of intelligence of the labour employed has no slight influence upon the value of mineral properties; for it is well known that where there are steady, and contentedly settled men, with which to carry on mining operations, there is a great advantage in comparison of that where men are unsettled and roving, and consequently less easily controlled and disciplined, and less amenable to reason.

When mining property comes into the market for sale, its real value is a matter not always easily determined. Foreign enterprises are occasionally offered to speculators and capitalists, and often consist of concessions held from foreign governments; these concessions we have known in some instances to have been both numerous, and widely dispersed; and instead of a large number of concessions being a source of strength to foreign mining it is much more frequently its weakness, for it is mostly made compulsory upon the holders that a certain amount of work shall be performed at each concession yearly, or the right to hold is endangered, and may be forfeited: and thus are many companies led into attempting too great a series of operations at the first, which, in new countries, and with untrained labour, often involve so heavy an expenditure that not a few collapse under the burden; while if attention had been confined to some two of the most promising and conveniently situate until they were fairly developed and made productive, much less preliminary cost would have been incurred, much capital saved, and the other concessions could then be gradually attended to and developed in a manner most conducive to the permanent success of the entire undertaking; and much fewer companies would then be brought to ruin.

If mining properties in foreign countries are much scattered there is much attendant expense in opening up and maintaining roads, or channels of communication and transit for the minerals; for, most likely, there will often be but few native roads that can be made available in any but detached parts; and where this is so, and native labour is used, though it may be available in sufficient abundance it may still be unacquainted with the class of work to be performed, and thus, while the labour is nominally cheap, it may in reality be very much dearer than a higher paid class of labour in other countries; and, as a consequence, the results of carrying on work in foreign countries seldom realize the expectations raised on the basis of cheap labour. These things will be elements in the calculation, in seeking to form an estimate of the value of foreign mining or other properties for the profitable employment of capital, and great care will be needful in making such valuations: for while doubtless there are many rich properties awaiting the application of capital, to yield a

most remunerative return, there are others from which little
can be properly expected.

The same may be said of properties in our own country,
coming into the market; for while there are some properties
which are thoroughly legitimate and sound, and offer a good
security for the investment of capital, there are some occasionally offered which, if they ever have paid a return on capital,
have now pretty well ceased to have any prospects of doing so
any more; in fact the owners may wish to sell them in consequence of their being of no value to hold. It is, however,
certain, that such transactions are not confined to mining, for
in many other ways we find the same principle acted on; and
if a person with his eyes open buys a worthless thing he has
himself principally to blame, if he did not take due precaution
to ascertain by the best available means whether it was of the
value represented to him.

While coal veins, and other stratified deposits, admit of
the application of general estimates of acreage yields of unworked ground, and with the cost at which the mineral can be
produced and sold supply general information for computation,
casual deposits cannot often have such principles wholly
applied to them; for they are in some cases very variable in
their features, and so irregular in the quantities yielded by
similar areas, that it is but the merest guess-work to assign
any quantity as a probable yield.

Doubtless, in some of the hematite iron ore deposits, the
further we follow them down the greater we may find their
bulk to be, and probably their richness also; but the manner
of working imposed by the nature of the deposit, and the
precautions necessary to ensure against flooding, will be
certain, in these deposits, to affect their value in the course of
time; and what will be a good system for one deposit in one
country, may in another country, and with precisely similar
surroundings of stratification etc., be quite inapplicable: for if
we suppose the case of a hematite deposit, very rich in itself,
but situate where the Lancashire practice of working downward would be inapplicable in consequence of semi-tropical or
violent rains prevailing at times—an instance of which we

have known, and have been connected with, and directing, in Portugal—some other mode of working must be adopted, so as to keep the surface intact; for were it in such cases to be broken by the subsidence brought about by working downward from the surface, the immense rains which in a very short time occasionally convert the surface of the ground into one wide sheet of water, would then quickly bring about the destruction of the mine, not alone by flooding the lower workings, but by carrying down into the workings such great quantities of the small refuse and sand overlying the district, that it would become a matter of *reopening* the mine to get at the former workings again. Of course, when the deposit being worked is situate on a hill, where good natural drainage can be arranged, such inconvenience from this cause does not arise, or is easily provided for. Under the circumstances we have mentioned, the only feasible plan seems to be that of going sufficiently deep to commence operations, and then to work upwards towards the surface, filling in the excavated space with rough material sent down into the mine; and always leaving a sufficient thickness untouched, to ensure an unbroken crust to keep out the flood waters.

These and similar circumstances will all go to tell upon the value of mineral undertakings, of the different kinds to which they respectively apply; so that the correct estimation of the value of a mineral property is often a complicated question, involving several matters which require special consideration,—and can only be duly arrived at by the examination of each property, not only upon general principles, but also with regard to the special features of the nature of the deposit; country in which situate; facilities for transit; abundance of labour; cost of establishing, developing, and conducting the operations; market for produce at a profitable price; and other matters of a relative nature which will suggest themselves to a competent valuer, who may have to entertain the question of the valuation of such properties.

We have found that veins of coal of thicknesses varying from 2 ft. 6 in. to 5 ft. have yielded the following different numbers of tons per foot of thickness per acre: viz. 1100,

1090, 920, 1300, and 1000 tons, statute weight—taking the exhausted ground as measurement; and these numbers will probably be found to be about what it is possible to realize in practice generally, though some of the quantities may be considered to be very low; but several causes go to influence the total quantity to be obtained from any particular vein,—not only the varying specific gravity of coals, but the strength or firmness of the veins, the nature of the superincumbent strata, mode of working, and also, in no slight degree, the system of management.

As being amongst the prominent questions that would have to be entertained and considered, in valuing any colliery that may be offered for sale, are some or most of the following.

1st. The area and compactness of the property, provisions and clauses of lease:—are they reasonably favourable to the Lessee, and what is the royalty reserved, dead rent, average clause, power of assignment, of removal of plant at expiration or other determination of lease etc. etc. etc. ?

2nd. What is its relative position in regard to other properties being worked ?

3rd. What is its relative position for market purposes for produce; to what markets can its produce be sent; and has it good railway, or other sufficient service ?

4th. What are the general known features of the strata in respect of Faults, Springs, Swamps, old Workings (if any) etc., and how situate ?

5th. Are any of the veins of mineral—coal or iron—in the same property, that can be worked to the detriment of that whose value is sought, demised to *other* lessees ?

6th. Quantity and quality of coal unworked, and its accessibility from Pit, or Level; and will the coal last out the term of the lease, and cover the dead rent to the end ?

7th. What is the price at which the coal can be delivered at "bank" in ordinary times, all working charges paid; and what are the royalties, wayleaves, and other charges binding on same ?

8th. What profit per ton will this cost admit of in ordinary states of the market; and what quantity is the colliery capable of producing per diem?

9th. Does the vein of coal give off much explosive gas; is the "top" good; and does it require much timber?

10th. Is there a good sound system of ventilation in operation, with sufficient air, and shafts etc. for same?

11th. How many years will the royalty last to *give profit?*

12th. Is the colliery well provided with machinery, plant etc., requisite for raising and despatching the coal; and are good plans of all the *old* workings obtainable?

13th. Are the boundaries infringed upon at any point; and do any liabilities attach to the property other than those usual and ordinary?

14th. What facilities of access to other veins, or royalties, does the possession of the present field and vein, or veins, give; and what is their money value to an incoming tenant or purchaser?

15th. What convenience, and area of spoil ground, is there for tipping rubbish etc. from the colliery?

16th. Do the minerals underlie any town, village, reservoir or accumulation of water, or any other notable place or thing, which would render the working of the coal liable to, or for, any damages, or actions at law, more than ordinary?

17th. Then:—Under all the circumstances what net profit per annum will the colliery produce, and for how many years, at a certain estimated output per annum?

An amount should be allowed off the profit for an insurance or compensation fund (in case of accident), and the income should provide a sufficient amount, after that deduction, to give a good interest on capital laid out, as well as to provide for the recouping of capital, before the colliery becomes exhausted.

The annual value etc. having been thus obtained, and any other matter observed necessary being duly considered, the present value will have to be worked out by the ordinary rules for money computations of this nature, and the result shewn in a proper tabulated form.

We will conclude by giving a few examples of the valuation
of the freehold interest in mineral property; and also of a reserved rent-charge: and these will also, to a considerable extent,
illustrate the principle to be followed in estimating the value
of a terminable income, annuity, or other interest, arising from
a colliery, or other property generally.

Reserved Rent Charge.

ExAMPLE I.

A person is entitled to an annual rent-charge of £140
upon a property, for a remaining term of 17 years; and it is
required to clear the property of this charge:—the proprietor
is willing to accept an immediate equivalent, and to allow
interest at the rate of 3½ per cent. per annum. To what
amount does he become immediately entitled?

To solve this, we require to find the value of an immediate annuity of £1 per annum, at that rate of interest;
this is found by Smart's Tables to be £12·65132, and if we
multiply this amount into the rent-charge we have 12·65132 ×
140 = £1771·184 for the value of the immediate equivalent.

We may also work out the same result by Thoman's Tables
—using logarithms.

Thus, taking same term and rate,

$$\text{Comp. log. } a^* = 1\cdot1021358$$
$$\text{Log. } 140 = 2\cdot1461280$$
$$\text{Log. } 1771\cdot184 = 3\cdot2482638 = \text{the same amount}$$

as that above given, or £1771. 3s. 8d.

- A little consideration will suffice to shew that these
formulæ will apply to other questions of money interests. For
instance, the computations will apply if the question is:—
What is the present value of an immediate annuity for 17
years, in any property, if it is required to make an interest
of 3½ per cent. per annum? for a person who is liable to
pay an annual rent-charge, if he rids himself of the liability

by paying down a certain sum, does, in effect, purchase to himself an annuity equal in amount to the rent-charge he was before paying. Other applications of the same principle will readily suggest themselves to the enquiring mind.

If the result of the computation be first obtained, while treating the annuity as being one single yearly payment, though it be, in fact, really divided into quarterly payments, or half-yearly payments, we may find the increased value under the latter circumstances very simply by Thoman's Tables.

EXAMPLE 2.

Question. A property yields a yearly rent of £860: what is it worth as a perpetuity when the rent is paid yearly; also when paid half-yearly; and when paid quarterly? Interest at the rate of $7\frac{1}{2}$ per cent. per annum.

1st. When paid yearly.

Comp. log. a^u $= 1\cdot1249387$
Log. 860 $= 2\cdot9344985$
Log. £11,466·66 $= \underline{4\cdot0594372}$ = the Value.

2ndly. When paid half-yearly.

Log. of 1st value $= 11,466\cdot66$ $= 4\cdot0594372$
$+ Q_2$ equivalent $= 0\cdot0080676$
Log. £11,681·664 $= \underline{4\cdot0675048}$ = the Value.

To prove the second computation, it is sufficient to shew that the 1st value is increased by an amount = to the value of a perpetuity of $\frac{1}{2}$ year's interest on $\frac{1}{2}$ the annuity, at $7\frac{1}{2}$ per cent. as before; viz. £16·125.

Then Comp. log. a^u $= 1\cdot1249387$
Log. 16·125 $= 1\cdot2074997$
Log. £215 $= \underline{2\cdot3324384}$ = the Value; being

the amount by which the 1st value is increased in the 2nd computation: viz. $11,466\cdot66 + 215 = £11,681\cdot66$.

3rdly. What is the value when the annuity is paid quarterly, the same rate of interest being allowed?

Log. of 1st value $= 4\text{·}0594372$
$+\ Q_4$ equivalent $=\ \text{·}0121951$

Log. £11,793·216 $= \overline{4\text{·}0716323}$ = the Value; being the value when the annuity is paid quarterly.

To prove the 3rd computation the proceeding is more intricate; but it may be shewn that the 1st value, £11,466·66, is increased by the perpetuity of the following amounts, viz.:

1st. $\frac{3}{4}$ interest on 1 quarter of the annuity
$= £12\text{·}09375 =$ in value £161·249

2nd. $\frac{1}{2}$ interest on 1 quarter of the annuity
$=\ £8\text{·}0625 =$ in value £107·500

3rd. $\frac{1}{4}$ interest on 1 quarter of the annuity
$=\ 4\text{·}03125 =$ in value £53·735

and by the compound interest on the quarter's interest on the quarter's interest on the quarters of the annuity $\Big\}$ $\ =\ $ £4·072

Then, Amount of Value when paid yearly $=$ £11,466·660

Total Value as per 3rd computation $=$ £11,793 216

EXAMPLE 3.

The Valuation of Freehold Property with Minerals immediately productive.

What is the Value of an Estate of 400 acres, the minerals of which are just beginning to be worked, the fair estimated output per annum of coal being 120,000 tons (= 400 tons per day for 300 days) at a royalty of, say, 10d. per ton; and the estimated duration of the same being 80 years? This would produce an annuity of £5000 for 80 years. The surface is being let, as farm land and rough pasturage, at £250 a year. Interest on capital to be allowed at 8 per cent. per annum; and capital to be recouped at $3\frac{1}{2}$ per cent. per annum; interest on purchase money of surface being at the rate of 4 per cent. per annum, and in perpetuity.

Then, the surface will thus be worth 25 years' purchase, or a total derived thus: 250 × 25 = 6,250 . . . . . £6,250.   0

The annuity of £5000 for 80 years, at 8 per cent. per annum, and to recoup at 3½ per cent. per annum, is worth in present money the sum total of £60,690. 15. 0

Total Value £66,940. 15. 0

The formula by which the second value is derived is that given by Thoman in his treatise, and is as follows :

$$V = \frac{a}{t' + \frac{a''}{r''}},$$

in which V is the value; a the annuity; t' the rate per cent. to be made on purchase money; a'' the annuity which £1 will purchase for that term, and at rate of recoupment; and r'' the amount of £1 at the end of the term, and at the same rate.

Then applying this formula to the foregoing case we have the following values :

Logarithms.

Log. a'' = 8·5726961

Log. r'' = 1·1952280   Nat. Numbers.

$a''r^{-n}$ = 7·3774681 = ... 0·0023845

+ t' = 0·08

Log. $t + \frac{a''}{r''}$ = 8·9158475 = ... = 0·0823845

Log. a = 5,000 = 3·6989700

Log. of V = 4·7831225 = £60,690·75, or £60,690. 15s. 0d.

= purchase money of the annuity.

We may check the above computation by working out the question by the rules given by Hardy as in Inwood's Tables at page 179.

1st. by Smart's Tab. 3, $(a)^{80}$ = 419·30678

5000

Product = 2096533·9

G

2nd. $(a)^{80} = 419\cdot30678$

$$\times t' = \underline{\quad\cdot08\quad}$$

Product $= 33\cdot5445424$

$$+ \underline{\quad I\quad}$$

Sum $= \underline{34\cdot5445424}$

3rd.

$$\frac{2096533\cdot9}{34\cdot5445424} = £60,690\cdot75 = V \text{ the value as before.}$$

Then, the interest on this value at 8 per cent. per annum will amount to................4th. .... $4855\cdot26$
giving for amortisement at $3\frac{1}{2}\frac{0}{0}(=s)$ $144\cdot74$

Total annuity $\underline{£5,000\cdot00}$

And to prove the sufficiency of the amortisement we have, by following the same rule,—

5th.   Smart's Tab. 1.

$$r^{80} = 15\cdot67573$$

$$- \underline{\quad1\cdot\quad}$$

Diff. $= \underline{14\cdot67573}$

6th.   Log. $14\cdot67573 = 1\cdot1666016$
         Log. $s = 144\cdot74 = 2\cdot1605886$
   C' Log. rate $\cdot035 = 11\cdot4559320$

     Log. $60,690\cdot71 = \underline{4\cdot7831222} =$ the capital recouped

at the end of the term of 80 years.

The foregoing estimate of quantity of coal would require a total thickness of 25 ft. or thereabouts, over the whole area of the property.

EXAMPLE 4.

Freehold Property with Minerals, being partly immediately productive, and partly deferred, and comprising the following lots, viz. :

Lot A.   Surface rental of 350 acres, equal to a perpetuity and let at a yearly sum of £145, net.

Lot B. The upper veins of Coal producing 200 tons per day, or 60,000 tons per annum, let at 9*d.* per ton royalty; and calculated to yield this quantity for a further term of 18 years.

Lot C. The lower veins of Coal which from their favourable position are likely to be productive in 15 years, and then to yield an output of 300 tons per day, or 90,000 tons per annum, for a term of 65 years, and to let at 1*s.* per ton royalty.

Lot D. Freehold groundrents from houses built, amounting to £200 per annum, for 75 years unexpired term of the leases.

Lot E. Royalties or wayleaves yielding, at 1*d.* per ton, £80 per annum; and estimated so to continue for a further term of 12 years.

Then supposing interest to be allowed on the several lots as follows, viz:—

Lot A. 4 per cent. per annum as a perpetuity,

Lot B. 8 per cent. per an. on Capital, and to recoup at $3\frac{1}{2}\frac{0}{0}$,

Lot C. 10 ,, ,, ,, ,, ,, ,, $3\frac{0}{0}$,

Lot D. 7 ,, ,, ,, ,, ,, ,, $3\frac{1}{2}\frac{0}{0}$,

Lot E. 10 ,, ,, ,, ,, ,, ,, $3\frac{0}{0}$,

the following will be the values: the same notation and rules being applied as in Example 3.

The Value of Lot A will then be $145 \times 25 = £3,625.$ 0*s.* 0*d.*

Lot B. Annuity for 18 years £2,250.

Interest on Capital 8 per cent.

Interest to recoup $3\frac{1}{2}$ per cent.

$$\text{Logarithms.}$$

$$\text{Log. } a^{18} = 8\cdot8797657$$

$$\text{Log. } r^{18} = 0\cdot2689262$$

Nat. Numbers.

$$\text{Log. } a^{18}r^{-18} = 8\cdot6108395 = \ldots\ldots\ldots\ldots 0\cdot0408168$$

$$+ t' = 0\cdot08$$

$$\text{Log. } t' + \frac{a^{18}}{r^{18}} = 9\cdot0821269 = \ldots\ldots\ldots\ldots 0\cdot1208168$$

$$\text{Log. } 2,250 = 3\cdot3521825$$

$$\text{Log. of Value} = 4\cdot2700556 = £18,623\cdot26.$$

Checking by Hardy's rules as in Ex. 3,

1st. Smart's Tab. 3, $(a)^{18} = $ 24·49969

$$2,250$$

Product $= 55124·3025$

2nd.

$$(a)^{18} = 24·49969$$
$$\times t' = ·08$$

Prod. $= 1·9599752$
$$+ 1·$$

Sum $= 2·9599752$

3rd. $\dfrac{55124·3025}{2·9599752} = 18,623·23 = $ Value.

4th. Int. on Value at 8 $\frac{0}{0}$ $= 1489·8584$
Amort. to recoup $3\frac{1}{2}$ $\frac{0}{0}$ $=$ 760·1416

The annuity $= £2,250·0000$

5th. Smart's Tab. 1. $r^{18} = 1·85749$
$$- 1·$$

Diff. $= 0·85749$

Then 6th. Log.   ·85749 = 9·9332291
Log. 760·1416 = 2·8808944
C' Log.   ·035 = 11·4559320
Log. 18,623·25 = $\underline{4·2700555}$ = the capital re-

produced.

Lot C.   The annuity for 65 years equals £4,500.
Interest on Capital $=$ 10 per cent.
Interest to recoup $=$ 3 per cent.
Period the annuity is deferred $=$ 15 years.

Logarithms.

$$= 8\text{·}5458736$$
$$-\ 0\text{·}8344196$$

Nat. Numbers.

$$\overline{7\text{·}7114540} = \ldots\ldots\ldots\ldots\ldots 0\text{·}0051458$$
$$+\ l' = 0\text{·}10$$

$$l' + \frac{a^{65}}{r^{65}} = 9\text{·}0217900 = \ldots\ldots\ldots \text{Sum} = \cdot1051458$$

$$4{,}500 = 3\text{·}6532125$$

Log. of $V = \underline{4\text{·}6314225} =$ Value, treated as "immediate," $= £42{,}797\text{·}9.$

Checked by Hardy's rules,

1st.   Smart's Tab. 3, $(a)^{65} = 194\text{·}3328$
$$\underline{4{,}500}$$
$$\text{Product} = \underline{874497\text{·}60}$$

2nd. $\qquad\qquad (a)^{65} = 194\text{·}3328$
$$\times\ l' = \qquad \cdot10$$
$$\text{Prod.} = \overline{19\text{·}43328}$$
$$+\ 1\cdot$$
$$\text{Sum} = \underline{20\text{·}43328}$$

3rd. $\dfrac{874497\text{·}6}{20\text{·}433280} = 42{,}797\text{·}71 =$ Value, as "immediate."

4th.   Int. on Value at $10\frac{8}{8} = 4279\text{·}771$
$\qquad$ Amort. to recoup at $3\frac{8}{8} = \ 220\text{·}229$
$\qquad$ Total annuity $\qquad\qquad \underline{£4{,}500\text{·}000}$

5th.   Smart's Tab. 1. $r^{65} = 6\text{·}830$
$$-\ 1\cdot$$
$$\text{Diff.} = \underline{5\text{·}830}$$

6th.   Log. $5 \cdot 83 = 0 \cdot 7656686$

        Log. $220 \cdot 229 = 2 \cdot 3428746$

        C^t Log. $\cdot 03 = 11 \cdot 5228787$

        Log. $42,797 \cdot 85 = \underline{4 \cdot 6314219}$ = the capital reproduced.

But as this annuity is deferred for 15 years (the first payment coming due in the 16th year) the above value has to be discounted for that term; and supposing that 8 per cent. per annum discount is allowed for it we have to proceed as follows. The interest at 8 per cent. per annum on £42,797·9 is £3,423·832.

Then (by Thoman's)   Log. $3,423 \cdot 832 = 3 \cdot 5345164$

           ÷ by Log. a^{15} $= 9 \cdot 0675527$

             Log. $29,374 \cdot$ $= \underline{4 \cdot 4669637}$

∴ $42,797 \cdot 9 - 29,374 = \underline{£13,423 \cdot 9}$ = the value of the deferred annuity of Lot C treated on the terms indicated.

Lot D.   The annuity for 75 years = £200.

        Interest on Capital   = 7 per cent.

        Interest to recoup   = 3½ per cent.

Logarithms.

Log. a^{75} $= 8 \cdot 5782860$

Log. r^{75} $= 1 \cdot 1205262$ Nat. Numbers.

Log. $a^{75} r^{-75} = \underline{7 \cdot 4577598}$ $= \dots\dots\dots\dots \cdot 0028692$

                              $+ t' = \cdot 07$

Log. $t' + \dfrac{a^{75}}{r^{75}} = 8 \cdot 8625440 = \dots\dots$ Sum $= \underline{\cdot 0728692}$

Log. 200 $= 2 \cdot 3010300$

Log. V $= \underline{3 \cdot 4384860} = \underline{£2,744 \cdot 64}$ = Value

By Hardy's rules,

1st. $(a)^{75} = 348 \cdot 53001$

                   200

     Product $= \underline{69706 \cdot 002}$

2nd.
$$(a)^{75} = 348\cdot53001$$
$$\times t' = \cdot07$$
$$\text{Prod.} = 24\cdot3971$$
$$+ 1\cdot$$
$$\text{Sum} = 25\cdot3971$$

3rd. $\dfrac{69,706\cdot002}{25\cdot3971} = £2,744\cdot64 = $ Value as above.

4th. Then Int. on Value at $7\frac{0}{0} = 192\cdot1248$
Amort. to recoup at $3\frac{1}{2}\frac{0}{0} = 7\cdot8752$
$$\text{Total annuity} = £200\cdot0000$$

5th. $r^{75} = 13\cdot19855$
$$- 1\cdot$$
$$\text{Diff.} = 12\cdot19855$$

6th. Log. $12\cdot19855 = 1\cdot0863082$
Log. $7\cdot8752 = 0\cdot8962616$
C' Log. $\cdot035 = 11\cdot4559320$
Log. $2,744\cdot74 = 3\cdot4385018 = $ the capital reproduced

at the end of 75 years.

Lot E. The annuity for 12 years equals £80.
Interest on Capital = 10 per cent.
Interest to recoup = 3 per cent.

Logarithms.
Log. $a^{12} = 9\cdot0020022$
Log. $r^{12} = 0\cdot1540467$ Nat. Numbers.
Log. $a^{12}r^{-12} = 8\cdot8479555 = \dots\dots\dots\dots = \cdot070462$
$$+ t' = \cdot10$$
Log. $t' + \dfrac{a^{12}}{r^{12}} = 9\cdot2316273 = \dots\dots$ Sum $= \cdot170462$

Log. $80 = 1\cdot9030900$

Log. $V = 2\cdot6714627 = £469\cdot313$ the Value.